HORSEGRAM IN INDIA

Editor

D. KUMAR

Central Arid Zone Research Institute,
Jodhpur - 342 001 (Raj.)

SCIENTIFIC PUBLISHERS (INDIA)

P.O. BOX 91 JODHPUR

Published on behalf of
INDIAN ARID LEGUMES SOCIETY

Published by:
PAWAN KUMAR
Scientific Publishers (India)
5-A, New Pali Road, P.O. Box 91
JODHPUR – 342 001
Tel.: +91-291-2433323
Fax.: +91-291-2512580
E-mail: info@scientificpub.com
www.scientificpub.com

ISBN: 81-7233-451-6

Lasertype Set: Rajesh Ojha

Printed in India

PREFACE

Scientifically less known and under investigated, horsegram is an important arid pulse known for exceptionally drought hardy nature in concomitant with little aftercare and agronomic inputs. The crop has thus, become poor men's crop owned by marginal farmers on degraded and resource constraint soils. The popularity of the crop can be settled from its suitability towards a variety of soils: shallow red soils, gravelly soils, hard uplands, red sandy soils, acidic lateric soils (pH 5.5), black cotton soils, clayey, paddy soils, stony soils, cleared rough forest etc; in many states. This drought hardy crop is grown in almost majority the states with rainfall range of 250-700 mm. But almost 95% area of this legume is confined in five states of Orissa, Tamil Nadu, Karnataka, Maharastra and Andhra Pradesh. Besides, being loaded with high seed protein fraction, it posses high amount of calcium, iron, phosphorus and vitamins contents required for vegetatarian diets. The crop is reckoned with for its medicinal uses, particularly as a cheap and steady treatment against stone formation in kidney and gal bladder and many other less known ayurvedic treatments are also associated with this arid legume.

The crop ideal for a number of double/ sequence/ rotations combinations in different climatic and geographic zones with annuals, perennials, grasses etc; has, therefore, proved vital crop component in cropping system. Horsegram, has therefore, occupied the position of cheapest source of livelihood of the poorest of poor resource ridden peasants and farmers of arid zones in India.

This annual legume is also known for some inherited defects and demerits. For instance, slow growth and long maturity period coinciding with terminal stress, higher production of biomass with poor conversion, susceptibility to anthracnose, powdery mildew and certain virus diseases, therefore, rendering it in poor productivity.

Photo and thermo-sensive nature of horsegram is biggest barrier in hampering expansion of its area, consequenting in poor adoption of promising strains across the zones and seasons. Research efforts attempted do not match with its large acreage, great social needs and uses in many ways. Thus, concerted efforts on initial technological advancements in horsegram are quite slow, therefore, depicting gloomy picture on its status of available research information. It is thus, desired that efforts are initiated to collect, compile and condense research results on this crop available in pieces and scattered at varied levels and sources. Present compilation on **Horsegram in India** is an attempt in this direction. The book is divided in ten chapters dealing with introduction, genetic improvement, crop production, quality aspects, plant protection and status of main horsegram growing states, contributed by the scientific personnels engaged in horsegram research in India.

It is hoped that the book in question will be of desired value and use to the scientists, scholars, administrators and planner engaged in improvement of horsegram growth in India in one way or other.

Place : Jodhpur
Date : May 15, 2006

D. KUMAR

ACKNOWLEDGEMENT

It always yields consistent pleasure and joy to me while writing any acknowledgement to all those who have ever rendered their help and support in bringing out any publication on arid legumes.

It is my foremost and immediate duty and desire to extend warmest gratitudes for helping and guiding hands extended by Dr. Mangala Rai, Secretary DARE & Director General, ICAR, New Delhi. His support in a variety of ways has been innumerable and unshapable for present publication and related acts and actions concerning Arid Legumes Network Project.

Rich words of thanks and innest desire of mine always sprut spontaneously while I think to thank Dr. Gautam Kalloo, DDG, (Crop Science and Horticulture). He has been sole motivator and real encourager for arid legumes research activities and also in bringing out this compilation on horsegram.

Dr. Pratap Narain, Director, CAZRI, Jodhpur has been appreciative as and when required. He has been constently watching, monitoring and guiding me and affairs related to this book and this project. He proved a source of inspiration. Therefore, I thank him.

Dr. N.B. Singh, ADG (Seed) and Dr. O.P. Dubey, ADG (O&P) brought me to the level of composing this book. They have always been source of mental promotion. I therefore, profusely thank them for their referred help. He was Dr. N.B. Singh who inspired and injected me to compile literature on Arid Legumes during arid legumes Workshop at S.K. Nagar Workshop in May 2000. Had he not reminded me regarding my conviction and dedication towards publication related to Arid Legumes probably, I would have never come across. Thus, lion share of acknowledge goes to Dr. N.B. Singh.

My special, thoughtful and spontaneous acknowledgement goes in favour of all the writes and the contributors who have contributed their enormously valued research information generated after burning mid night oils. The name for instance can be mentioned of Dr. S.B.S. Tikka, Director of Research, Dr. S.D. Solanki, ARO, Dr. N.B. Patel, and Dr. R.M. Chouhan Professor, S.D. Agril. Univ. S.K. Nagar; Dr. H.S. Gupta, Director, Dr. V. Mahajan, Head of the Division and Dr. S.K. Shukla, Scientist, Vivekanand Parvartiya Anusandhan Sansthan, Almora; Dr. A. Henery, Principal Scientsit (Plant Breeding) CAZRI, Jodhpur; Dr. K.P. Vishwanatha, Professor (Plant Breeding), GKVK, Bangalore; Dr. S.K. Sharma, Dean, College of Basic Sciences, Dr. H.L. Thakur, ADR, Dhaula Kuan, Mrs. Daisy Basandrai and Ashwani K. Basandria from CSKHPKV, Palampur; Dr. S.M. Purusothaman, Asstt. Scientist (Plant Pathology). Mani Chellappan and K. Karthikeyan, KAU, Pattambi (Kerala) and . Dr. S.N. Sodani, Arjia, Bhilwara.

I may not forget mentioning the names of my colleagures in PC Unit, helping and shaping this compilation in present form. They have always stood with me in referred regard. Mention can be made of Mrs. Arti Shekhawat, Mr. P.S. Bhati, and Dr. H.R. Mahala, Miss Deepa and Mr. J. Gaur.

In the end, I express my deep sense of willingness to feel obliged in writing a few lines for typing the script of this book many times by Mr. Surender Bhati. Name can also be mentioned of Mr. Harish Chandra Sharma for typing the script and giving final touch to it at the end.

Place: Jodhpur **(D. Kumar)**

CONTRIBUTORS

Abraham, Z. Head, NBPGR, Regional Research Station, Vellanikkara, Thrissur - 680 656, Kerala

Asha, K.I. Scientist (Sr. scale), Econ-Botany, NBPGR, Regional Research Station, Vellanikkara, Thrissur-680656 (Kerala)

Ashwani, K. Basandari Asstt. Scientist, Department of Plant Breeding, Chaudhary Swaran Kumar Himachal Pradesh Krishi Viswa Vidhyalaya, Regional Research Station, Dhaulakuan (Sirmore) 173 001 (H.P.)

Chauhan, R.M. Professor, Dept. of Plant Breeding, College of Agriculture, S.D. Agricultural University, S.K. Nagar, (Gujarat).

Chellappan, Mani Kerala Agricultural University, College of Horticulture, Vellanikkara, Thrissure 680 656 (Ker.).

Daisy, Basandari Asst. Scientist, Department of Plant Breeding, Chaudhary Swaran Kumar Himachal Pradesh Krishi Vishwa Vidyalaya, Regional Research Station, Dhaulakuan (Sirmore) 173 001 (H.P.)

Gupta, H.S. Director, Vivekananda Parvatiya Anusandhan Sansthan, Almora, (Uttranchal).

Henry, A. Principal Scientists (Plant Breeding), Central Arid Zone Research Institute, Jodhpur - 342 003 (Raj.).

Jayan, P.K. P.S. (Retd.), Econ.-Bot. NBPGR, Regional Research Station, Vellanikkara, Thrissur-680656 (Kerala)

Karthikeyan, K. Kerala Agricultural University, Regional Agricultural Research Station, Pattambi - 679 306 (Ker.).

Kumar, D. Principal Scientist (Plant Breeding) Central Arid Zone Research Institute, Jodhpur - 342 003 (Raj.)

Latha, M. Scientist (Sr. scale), Plant Breeding, NBPGR, Regional Research Station, Vellanikkara, Thrissur-680 656 (Kerala)

Mahajan, V. Principal Scientists and Head, Division of Crop Improvement, Vivekananda Parvatiya Anusandhan Sansthan, Almora, (Uttranchal).

Maya, C. Nair Research Associate, NBPGR, Regional Research Station, Vellanikkara, Thrissur-680 656 (Kerala)

Misra, S.K. Head, Germplasm Evaluation Division, NBPGR, Pusa - 110 012 (New Delhi)

Patel, N.B. Asstt. Research Scientists, Main Pulses Research Station, S.D. Agricultural University, S.K.Nagar (Gujarat)

Purushothaman, S.M., Asst. Professor (Plant Pathology), Kerala Agricultural University, Regional Agricultural Research Station, Pattambi - 679 306 (Ker.).

Reddy, P. Ragu Ram Principal Scientist (Plant Breeding) Central Research Institutes for Dryland Agriculture, Santoshnagar, Hyderabad, (A.P.)

Sharma, S.K. Dean, College of Basic Sciences, Chaudhary Swaran Kumar Himachal Pradesh Krishi Vishwa Vidyalaya, Palampur, 176 062, H.P.

Shukla, S.K. Scientist (Plant Breeding), Crop Improvement Division, Vivekanada Parvatiya Anusandhan Sansthan, Almora, (Uttranchal)

Sodani, S.N. Professor (Plant Breeding), Maharana Pratap Agricultural University, Regional Research Station, Bhilwara, (Rajasthan).

Solanki, S.D. Asstt. Research Scientists (Plant Breeding), Main Pulses Research Station, S.D. Agricultural University, S.K. Nagar, (Gujarat).

Sreenivasan, E. Associate Professor (Bot.) Kerala Agricultural University, Regional Agricultural Research Station, Pattambi - 679 306, (Kerala)

Thakur, H.L. Addl. Director of Research, Chaudhary Swaran Kumar Himachal Pradesh Krishi Viswavidyalaya, Reg. Research Station, Dhaulakuan, (Sirmore) 173 001 (H.P.)

Tikka, S.B.S. Director Research, S.D. Agricultural University, S.K. Nagar, (Gujarat).

Viswanatha, K.P. Professor (Plant Breeding), College of Agriculture, University of Agricultural Sciences, G.K.V.K., Bangalore, (Karnataka)

CONTENTS

1

Horsegram Research - An Introduction

D. Kumar

Horsegram commonly known as "Kulthi" is an important Arid Legume crops of India. It is a traditional tropical grain legume, well known for its hardiness and adaptability to poor soils and adverse climatic conditions that may prove unsuitable for most the other crops. In many parts of South India, horsegram is grown as a preparatory crop in newly reclaimed lands to improve the soil fertility by fixing nitrogen and increasing organic matter status through shedding of leaves (Sen and Bhowal, 1959). Horsegram is extensively grown in peninsular India and is cultivated up to 5000 feet elevation in Himachal Pradesh and Nepal. It is cultivated both in Kharif and Rabi seasons. At present, it ranks third among Indian pulses in area, covering about 1.7 million hectares with a total production of 0.74 million tons of the total cultivated area under pulses in India. The productivity of the crop is rather poor. The national average productivity of horsegram is 494 kg/ha against 539 kg/ha for grain legumes as a whole. This hardy legume thrives well in almost all types of soil, except highly alkaline soils. Light sandy soils are the best for this crop. It is also grown extensively on red loams, black cotton soils, and stony and gravely upland soils of Deccan, without much preparatory tillage. However recent surveys showed that its cultivation has been drastically reduced indicating its replacement by other crops. The Southern States are the major horsegram growing states in the country. About 93 per cent of the total area under horsegram in the country is occupied by six states viz., Karnataka, Maharashtra, Orissa, Andhra Pradesh, Madhya Pradesh and Tamil Nadu. In Karnataka about 34 per cent of the area under grain legumes is set for sowing horsegram. In Tamil Nadu the area under horsegram in about 18 per cent and in Maharashtra, Orissa, Andhra Pradesh and Madhya Pradesh, it is cultivated in about 9 per cent of the total grain legumes growing area (Table 1).

Table 1. Area, Production and productivity of horsegram in different states of India.

States	Area (000 ha)	Production (000ha)	Productivity (kg/ha)
Andhra Pradesh	272.5	76.3	280
Bihar	80.0	34.8	435
Himachal Pradesh	2.6	0.8	308
Jammu and Kashmir	2.1	0.6	286
Kerala	16.9	12.5	740
Karnataka	590.0	215.0	370
Madhya Pradesh	18.9	58.2	308
Maharashtra	301.3	94.5	314
Orissa	273.8	160.8	587
Tamilnadu	139.2	63.0	453
West Bengal	5.6	3.0	536
All India	1702.9	719.5	423

India produces about 0.81 million tones of horsegram annually, which constitutes about 7 per cent of total grain legume production in the country. However, the contribution of horsegram to the total grain legume production is about 28 per cent in Karnataka, 18 per cent in Tamil Nadu and around 10 per cent in Maharashtra, Orissa, Andhra Pradesh and Madhya Pradesh, each.

Uses

Horsegram is mostly utilized as animal feed, but it is also used as a human food for a large section of the rural population and is consumed as whole seeds, sprouts or whole meal alongwith cereal flour. The use of dry seeds of horsegram is very sporadic due to its poor cooking quality. This legume is said to have several medicinal properties for treating cough and bronchitis, kidney trouble and irregular periodicity of menstrual cycle. Although not valued so much as the other pulses it is equally nutritive. Being a cheaper pulse it is largely consumed by the poor people. South India is known for various tasty preparations out of this legume, such as *curry, pappad* and so on. *Panch dhani* a mixture of five crops (horsegram + Indian bean + cowpea + niger + castor) is grown by Karnataka farmers which is a common practice to combat drought and use horsegram and other crops.

Anti-nutritional factors

Oligosaccharides (raffinose, stachyose and verbascose) present in seeds are involved in flatulence production in human-beings and animals. However, there is reduction in these oligosaccharides on soaking of seeds. Antinutritional factors present in horsegram are: trypsin inhibitor, haemagglutinins, phytic acid and α-amylase inhibitor. However, on soaking and cooking, the contents of these antinutritional factors are significantly reduced. Dehusking results in marked improvement in protein efficiency ratio (PER) and digestibility, and also reduces cooking time from 2 hrs to about 30 min.

Medicinal value

Horsegram is known for medicinal (diuretic) properties (Watt, 1972) and is said to be good for patients suffering from urinary and kidney problems (Thakur, 1979). It is supposed to have medicinal value especially in dissolving stones developed in the kidneys. By consuming the soup prepared from sprouted horsegram, one can get better sleep. In ladies irregular monthly periods and urinary problems can be over come by consuming the grains daily. Horsegram pappad will reduce acidity. Grains are also useful in curing whooping cough. Constipation can be avoided by drinking a cup of horsegram soup (grain boiled water) with a small quantity of lemon juice and salt.

Nutrition

Horsegram is one of the cheapest sources of protein for both human beings and animals. The grain contains 22-24 per cent protein. However, a large variability in protein content of horsegram ranging from 18.5-28.5 with average 23.1 per cent has been reported (Mushtari Begum *et al.*, 1977). On an average, horsegram contains 57.2% carbohydrates, 0.5% fat and 3.2% minerals. The seed coat accounts for 13.7% of whole seed. It is also a cheap source of vitamins, calcium and iron (Katiyar, 1984). Horsegram is a rich source of urease.

The sulphur containing aminoacids cystine and methionine account for the lowest scores. The presence of Tripsin inhibitor has been reported. In addition, there is evidence that horsegram seeds contain growth inhibiting factors, which can be inactivated by autoclaving (Ray, 1969). The slow digestion of horsegram is attributed to the presence of Tripsin inhibitor (Sohonie and Bhandarkar, 1954).

Origin and Distribution

Horsegram [*Macrotyloma uniflorum* (Lam.) Verdc.] is distributed through out the tropics and is reported to be cultivated in India, Myanmar, Nepal, Malaya, Mauritius, Sierra Leone, Transval and West Indies; and grows wild in the eucalyptus forests of Queensland (Sundararaj and Thulasidas, 1986). It has been under cultivation in India since pre-historic times and according to Vavilov (1951) India is its primary centre of origin. But Dana (1976) reported that out of 240 species that are related to horsegram, only 23 are found in India and the remaining ones are located in Africa, therefore, Dana claimed that Africa is the primary centre of origin of this crop.

According to recent classification horsegram belongs to the subfamily Faboidae of family Fabaceae. Until recently it was considered under the genus *Dolichos.* However, Verdcourt (1982) studied the species in detail and found that it does not belong to the genus *Dolichos* but grouped under a distinct genus viz., *Macrotyloma* and named it as *Macrotyloma uniflorum.* One more species viz., *Macrotyloma axillare* identified under this genus is a wild relative and generally grown for fodder purpose.

Botanical Description

The species are tropical annuals or perennial herbs, usually prostrate or partly erect and normally having long runners. Horsegram is an annual succulent herb; slender, downy; slightly twining branches; springing from the base of the plants; semi-erect; low growing habit; having 30-50 cm height. Leaves are trifoliate; leaflets membranous entire, pilose, ovate, acute, 2.5-5.0 cm in length; stipulses 1 cm long, ovate lanceolate, minute stipules. Flowers are produced in leaf axils (1 3); peduncles are short; bracteate, bractealate; pedicellate, bisexuals, hypogynous, zygomorphic, complete, pentamerous, calyx downy, teeth lanceolate, corolla light yellow, papilonoceous, petals five standard longer than wings, scarcely reflexed, keel obtuse, narrow; stamens diadelphous (9+1); filaments alternately short and long anthers dithecous, introse, uniform, dorsified. Gynaecium with superior ovary, ovary monocarpellary, unilocular, with five to seven on marginal placentation. Style file form terminal, curved, stigma capitate and hairy around glabrous, fruit legume, flat. Pods linear recurved beaked, downy, dehiscent, 5-7 seeded. Seeds are small non-endospermous, flattened, rhomboidal, 3-6 mm long, light red, brown, black or mottled, testa shiny, hilum small and inconspicuous. Horsegram is self fertilized and matures in about 3-4.5 months, depending upon the variety.

Macrotyloma axillare, Macrotyloma africanum and *Macrotyloma daltonii* are three related species reported from Australia, are forage types. *Macrotyloma axillare* is known to be resistant to yellow mosaic disease and has been used to transfer this character to *Macrotyloma uniflorum.*

Chromosome number

Horsegram has the diploid chromosome number of 20. The chromosome evolution of *M. uinflorum* (horsegram) is believed to have progressed in two directions, one group having 20 small chromosomes (*M. uniflorum, M. baumani* and *M. axillare*) and the other having 22 large chromosomes (*M. glabrescence, M. lignosus* and *M. argentinus*). The intermediate forms have perennial tuberous roots and annual stems (*D. bulbosus*). Species relationships have, however, not been thoroughly understood.

Quantitative Genetics

Quantitative genetic studies as reported by several workers indicated that GCV prevailed for many of the yield associated characters. Nevertheless, the character number of pods per plant has shown the highest correlation with the yield.

Improvement in this crop to date is mainly through single plant selection and this has resulted in the development of a few improved varieties suited to local conditions. For instance, in Bombay three selections were found superior to the local check. In Madras state, culture No. 35, 5-122 and 123 were superior to local bulks. In Orissa, selections 27, 28, 39 and 164 were of promise. In Hyderabad, selections 33, 34, 36 and 42 were superior to check. In erstwhile Mysore state, a black Kulthi was found promising. The other promising selections were 8-1-1-8, Belgaum 1-8-3, Bijapur 1-6-5 and Hyderabad A-3-2-3. Subsequently BGM-1, which was a selection from Bailhongal (Belgaum District) was found stable and released as KBH-1 for the general cultivation by University of Agricultural Sciences, Bangalore. Recently another line PHG-9 which is a selection from Palampur from Himachal Pradesh was also released by the CVRC.

Planned hybridization followed by selection for several important characters have been at a low pace. However, since the crop is included in ICAR co-ordinated project on Arid Legumes, there is a ray of hope that the breeders in addition to individual plant selection, the other types of breeding is initiated in this crop as it has medicinal use of addition to being a consumable grain for both human and animals.

The most important disease prevailing on this crop in Karnataka is powdery mildew followed by YMV. The scientists should work in this direction very seriously since the powdery mildew not only reduces the yield to an extent to 70-80 per cent but even the 20-30 per cent of the grains obtained are of poor quality. Varieties with low growth inhibitors and tripsin inhibitors may find priority. Further, since grain has several medicinal value, the biotechnologists should take up work to manufacture some medicines especially for urinary problems after thorough understanding the chemical nature of this grains. The notion that it is a poor man's crop should go, on the other hand, the grain should be popularized by highlighting the medicinal value of this crop.

Since the crop improvement and productivity is due to convergent of several disciplines, the agro-nomists should develop a good package of practices, reduce the pest infestation in the field and thereby, in store. Economic control measures against diseases, proper storage methods against pests may also be found out. A good extension work with proven technologies to take up for horsegram cultivation.

Constraints to Higher Productivity

1. The improved varieties being cultivated have basically been developed following pure line selection from the local germplasm. Thus, conventional breeding methodology needs to be taken up for over all improvement in yield, quality and shorter growth period for adaptation.
2. The cultivation of local land races with long maturity expose this legume towards considerable moisture stress, during grain filling stage (terminal stress) resulting in low yield.
3. The pod shattering habit in local land races also causes considerable yield losses.
4. It is generally grown in poor soils with minimum agronomic inputs.
5. The crop is grown as second priority of the farmers on sub-marginal lands.
6. The role of this legume in various farming systems prevalent in its area of cultivation needs to be properly understood and worked out.

7. Horsegram is reported to be comparatively free from the attack of the pests and diseases. However, in certain agro-ecological ninches these disease-pests complex become important and cause considerable damage to the crop.
8. It is also attacked by storage pests (*Callosobruchus* sps.).
9. The technology developed for increasing the yield of this crop has not been disseminated to cultivators for achieving higher yields.
10. Photo and thermo-sensitive nature of horsegram does not permit its horizontal expansion in non-traditional and remote regions.

References

Dana, S., 1976. Origin, evolution and distribution of some grain legumes. *Indian J. Genet.,* 36: 143-145.

Katiyar, R.P. 1984. 'Kulthi' a promising pulse crop for Himachal Hills. Indian farming, XXXIV (9): 31-35.

Mushtari Begum, Srihara, J. and Hiremath, S.R. 1977. Varietal difference in protein of horsegram (*Dolichos biflours* Linn.). *Mysore. J. Agric. Sci.* 11: 521-524.

Ray, P.K. 1969. Alternative value of horsegram (*Dolichus biflorous*) Part I-Effect of feeding raw and treated seed flour on the growth rate of rats. *J. Nut. and Dietet,* 6: 329.

Sen, N.K. and J.G. Bhowal. 1959. Genetic studies in horsegram. *Indian J. Gen. Pl. Br.,* 19 (2): 228-233.

Sohonie, K. and Bhandarkar A.P., 1954. Quoted from Lata Manage and Kamala Sohonie. *J. Food Sci. & Tech.,* 9: 35-36.

Sundar Raj, and Thulasidas, G., 1986. Botany of field crops. The Macmillan company of India, Delhi.

Thakur, 1979. Scientific Crop Production Vol. 1. (Cereals and Pulses) Metropolitan book Co. Pvt. Ltd., New Delhi – 110 002.

Vavilov, N.L. 1951. The origin, ammunity and breeding of cultivated plants. *Ghronica Bot.* 13: 1-364.

Verdcourt, B. 1982. A revision of Macrotyloma (Leguminosae). Hooker's Icones. *Plantarum* 38: 1-138.

Watt, G. 1972. Dictionary of the Economic products of India. Gordhan and Co., New Delhi, pp: 175-183.

2

Genetic Resources

K.I. Asha, M. Latha, Z. Abraham, P.K. Jayan, Maya C. Nair and S.K. Mishra[1]

Horsegram or Madrasgram, *Macrotyloma uniflorum* (Lam.) Verdc. (Syn: *Dolichos uniflorus* Lam., *Dolichos biflorus* auct. non L.) belonging to the family Fabaceae is an important drought resistant dual purpose crop cultivated mainly in the tropics. The common name originates from its use as an important feed for horses and is equally important for human also. In India, its cultivation is known since prehistoric times. Domestication took place obviously here too. The Indian wild populations of horsegram correspond well with the cultivated ones. In breeding programmes the crop needs to be improved not only as a pulse, but also as a forage plant (Hanet, 2001). The crop being a native of India, is cultivated mainly in dry regions of Asia and Africa. In Asia, the crop is mainly cultivated in India, Malaysia, Myanmar, Nepal, Mauritius and West Indies. In India, is widely grown in Kerala, Tamil Nadu, Karnataka, Andhra Pradesh, Maharashtra, Bihar, Orissa and Madhya Pradesh. The availability of genetic diversity is rich in India and it is the only country cultivating horsegram in large acreages. It is used as a poor man's pulse in southern parts of the country where the seeds are consumed after boiling or frying. It is cultivated in areas with moderate rainfall exceeding 35 mm and is a very hardy and drought resistant crop suited to all soil types (Purseglove, 1974). Horsegram was mainly raised as a mono-crop in rice fallows of Kerala and presently restricted to the sub-drier regions, coastal Karnataka and Maharashtra and as a mixed one in interior Maharashtra, Karnataka plateau, Andhra Pradesh and Tamil Nadu. Though its productivity level is low, the crop is highly suited for low input agriculture in marginal environments (Patel *et al.,* 1995). Though considerable studies have been conducted on various aspects of the crop by several workers (Joshi *et al.,* 1994; Savithramma *et al.,* 1990; Sharma, 1995; Patil *et al.,* 1994; Dobhal and Rana, 1994a; 1994b; Lad *et al.,* 1999; Nagaraja *et al.,* 1999; Tripathi, 1999 and Jayan and Maya C Nair, 2001), its productivity level pitiably falls much below the sustainability line and gives immense scope for its improvement. It is cultivated under rainfed conditions either in the main season in Karnataka (June - October) and in Tamil Nadu (November - February) or as a second crop in Kerala (September - December) under the conventional cropping system. In Kerala, its dal is believed to be good for rheumatic patients and unlike other pulses *viz.,* pigeon pea, it is considered to be physiologically more useful and preparations in combination with other house hold but non-conventional vegetables such as banana rhizome, pseudostem, male flower bud etc. are made and consumed with rice as side dishes.

Wild forms of horsegram have also been reported in the Western Ghats especially in the Wildlife Sancturies. *Macrotyloma ciliatum* (Willd.) Verdc. is found in Tamil Nadu (Nair and Henry, 1983; Matthew, 1983) and

Andhra Pradesh (Pullaiah and Chennaiah, 1997). *Macrotyloma sar-garhwalensis* Gaur & Dangwal is a wild relative of horsegram found in the Garhwal Himalayas (Gaur and Dangwal, 1997). It is a non-twining annual herb with a protein content of 38.35 % which can be utilized in the breeding programmes for improvement of protein content. The protein content in cultivated horsegram is reported to be 16.9-30.4% (Patel *et al.*, 1995). Looking into the importance of this crop, its augmentation, evaluation and conservation are essential. Considerable genetic erosion started in the early 60s due to changes in cropping pattern and induction of new crops in the farming systems in India. Hence, collection of this important legume under the PL 480 scheme in the past and under the national plant exploration programme later by the Regional Stations of National Bureau of Plant Genetic Resources at Akola and Thrissur commenced in the late seventies. An effort has been made to present the status of horsegram genetic resources for their utilization in breeding programmes.

Exploration and collection

The Regional Station of NBPGR at Thrissur has made an earnest attempt to collect, evaluate, document and conserve the genetic resources of this crop since 1978. Under different exploration and collection programmes, a total of 1462 accessions could be accumulated. However, at present 1040 collections are being maintained. The well known local landraces collected from Tamil Nadu included *chembakanam, kanam, kanampayar, karumkanam, karupukanam, kollu, kothukollu, nattukollu, perunkollu, redkanam, vellakanam, vellakollu* and *whitekanam* and the one from Kerala included *kanam* from Idukki district of Kerala. Out of the 763 surviving accessions collected from 16 different states of India, 551 have been conserved on long term basis in the National Genebank, NBPGR, New Delhi (Table 1).

Table 1. Present Status of Genetic resomces of horsegram

Sl. No	State from which germplasm collected	No. of accessions		
		Collected	Surviving*	Conserved **
1	Andhra Pradesh	23	20	19
2	Bihar	41	38	26
3	Goa	10	8	7
4	Gujarat	7	7	6
5	Himachal Pradesh	29	28	22
6	Karnataka	136	69	40
7	Kerala	49	39	30
8	Madhya Pradesh	231	203	147
9	Maharashtra	83	72	57
10	Orissa	2	2	2
11	Punjab	6	6	3
12	Sikkim	6	6	6
13	Tamil Nadu	316	254	179
14	Uttar Pradesh	8	7	6
15	Uttaranchal	1	1	
16	Rajasthan	3	3	-
	Total	951	763	551

*8°C in the medium-term storage facility at NBPGR RS, Thrissur

** -20 °C in the seed store of National Genebank at NBPGR, New Delhi

Germplasm characterisation and evaluation

Out of a total of 1462 accessions of horsegram germplasm amassed, 1036 accessions were subjected to preliminary characterization during 1999-2004. A total of 965 accessions of horsegram collected from different agro-ecological situations of India was grown during *rabi* seasons of 1999-2004. The data were recorded on 12 qualitative and ten quantitative traits. Simple correlation and path analysis (Dewey and Lu, 1959) were done for different traits. Frequency distribution for all the quantitative characters was also carried-out.

Based on the observations on days to maturity, 965 accessions could be grouped in to 3 major categories, *viz.*, short duration type (<90 days for maturity), medium duration type (took 90-110 days) and long duration type (>110 days). Thus, 570 accessions belonged to short duration type, 260 to medium duration and 135 to long duration types. Critical observation on yield and other yield components in general showed that all the promising lines with higher seed yield belonged to long duration type. The seed yield per plant ranged from 0.22 to 7.31 g in short duration type, from 0.27 to 7.07 g in medium duration ones and from 0.21 to 11.86 g in long duration type. Variability observed in qualitative characteristics revealed that growth habit ranged between erect to very viny types, leafiness between sparse to abundant, leaf pubescence from puberulant to densely pubescent and stem colour between green and purple. Range in variability was maximum in pod colour and seed colour. Mature pod colour varied from straw, tan, cream, light brown, brown, dark brown, to brownish black. Variability in seed colour ranged from cream to black. The frequency distribution of the accessions in different qualitative characters is given in Table 2.

Variability studied for 10 quantitative traits showed significant range of variation in most of the traits (Table 3). The plant height ranged between 17.0 to 103.7 cm and with this primary branches per plant also varied from 1.0 - 9.8. With respect to this, yield components were very high in pods per plant (4 – 118), pod length (3.07 – 6.17 cm), 100-seed weight (0.927 – 4.108 g) and per plant yield (0.214 – 11.860 g). The range, mean, standard deviation and coefficient of variation with respect to all the traits are given in Table 3. The value of coefficient of variation was maximum for yield per plant (62.584) followed by primary branch number (59.218) and pods per plant (57.326), indicating higher range of variability in these traits.

Table 2. Frequency distribution of horsegram accessions

Sl. No.	Characters	Frequency distribution in qualitative traits – number of accessions													
		Character states													
		1	2	3	4	5	6	7	8	9	10	11	12	13	14
1	Growth habit	295	448	222											
2	Leafiness	206	337	422											
3	Leaf pubescence		15	463	487										
4	Branching	312	357	296											
5	Twining tendency	211	411	220	123										
6	Primary branching Pattern	78	321	419	147										
7	Stem base colour		2	963											
8	Stem top colour	374	210	381											
9	Stem pubescence		1	510	454										
10	Mature pod colour	44	54		216	389	244	10	8						
11	Seed colour		33	1			1	127	78	542	2	2	129	1	49
12	Seed mottleness	139	826												

1. Growth habit: 1. Erect 2. Viny 3. Very viny

2. Leafiness: 1. Sparse 2. Intermediate 3. Abundant

3. Leaf pubescence: 2. Puberulent 3. Moderately pubescent 4. Densely pubescent

4. Branching: 1. High 2. Medium 3. Low

5. Twining tendency: 1. None 2. Slight 3. Intermediate 4. Pronounce

6. Primary branching pattern: 1. Top 2. Top middle 3. Middle 4. Base

7. Stem base colour: 2. Light purple 3. Green
8. Stem top colour: 1. Purple 2. Light purple 3. Green
9. Stem pubescence: 2.Puberulent 3. Moderately pubescent 4. Highly pubescent
10. Mature pod colour: 1. Straw 2. Tan 4. Cream 5. Light brown 6. Brown 7. Dark brown 8. Brown and black
11. Seed colour: 2. Cream 3. Light brown 6. Grey 7. Mottled grey 8. Mottled brown 9. Mottled cream 10. Greenish brown 11. Chocolate 12. Black 13. Green yellow 14. Mottled greyish cream.
12. Seed mottleness: 1. Absent 2. Present

Table 3. Variability of different quantitative traits in horsegram germplasm

Sl. No.	Quantitative traits	Range	Mean	SD	PCV (%)
1	Primary branch number	1.0-9.8	2.663	1.577	59.218
2	Plant height (cm)	17.0-103.7	50.07	19.846	39.636
3	Pod length (cm)	3.07-6.17	4.639	0.561	12.093
4	Pod width (cm)	0.4- 1.0	0.598	0.078	13.043
5	Pods per plant	4.0- 118.0	25.524	14.632	57.326
6	Seeds per pod	3.4- 7.33	5.398	0.635	11.763
7	Seed length (mm)	3.58- 7.17	5.683	0.364	6.405
8	Seed width (mm)	2.63 – 5.12	3.924	0.291	7.415
9	100 seed weight (g)	0.927 – 4.108	3.023	0.339	11.214
10	Yield per plant (g)	0.214 – 11.860	2.507	1.569	62.584

Analysis of simple correlation was carried out in 45 combinations involving 10 productive traits. The results (Table 4) revealed maximum correlation between pods per plant and yield (79.1%), pod length and width (79.3%), plant height and pods per plant (71.2%) and primary branch number and pods per plant (69.4%). The correlation was negative between pod length and seed width and between pod width and seed length.

Table 4. Correlation among different characters in horsegram

Ch.	1	2	3	4	5	6	7	8	9	10
1	1.000									
2	0.5397	1.000								
3	0.0780	0.0732	1.000							
4	0.1725	0.0985	0.7932	1.000						
5	0.6941	0.7121	0.0489	0.1163	1.000					
6	0.1423	0.2562	0.5571	0.3864	0.2430	1.000				
7	0.1154	0.1754	-0.0519	-0.0530	0.1152	0.0035	1.000			
8	0.1385	0.2470	0.3356	0.2952	0.1618	0.2311	0.2817	1.000		
9	0.2217	0.3346	0.1065	0.1556	0.2598	0.0879	0.4354	0.4086	1.000	
10	0.4857	0.6335	0.1438	0.1115	0.7912	0.3292	0.1290	0.2269	0.2996	1.000

Note: The serial number of characters are given in Table 3.

Analysis of direct and indirect effects

Path coefficient analysis (Table 5) revealed that pods per plant had maximum direct effect (0.7670) on yield followed by pod length (0.1537). The direct effects were negative for primary branch number, pod width and seed length. This suggests the low value for selection of these traits. Maximum positive indirect effects are (0.5462) exerted by plant height and pods per plant followed by primary branch number and pods per plant (0.5324). Increase in plant height indirectly affects the yield by the increase in number of primary branches and number of pod bearing nodes. The low residual effect of 0.3228 obtained in the study indicates the adequacy and legitimacy of the characters used. Path co-efficient analysis revealed that branches per plant and seeds per pod had highest direct effect on grain yield which suggested that improvement in grain yield could be effective, if the selection is based on these characters (Pandya *et al.* 2003)

Table 5. Matrix of direct and indirect effects

T*	1	2	3	4	5	6	7	8	9	Correlation with yield
1	-0.1212	0.0528	0.0120	-0.0254	0.5324	0.0129	-0.0010	0.0048	0.0183	0.4857**
2	-0.0654	0.0979	0.0113	-0.0145	0.5462	0.0233	-0.0015	0.0086	0.0276	0.6335**
3	-0.0095	0.0072	0.1537	-0.1167	0.0375	0.0507	0.0004	0.0117	0.0088	0.1438**
4	-0.0209	0.0096	0.1219	-0.1471	0.0892	0.0351	0.0004	0.0103	0.0128	0.1115*
5	-0.0841	0.0697	0.0075	-0.0171	0.7670	0.0221	-0.0010	0.0056	0.0214	0.7912**
6	-0.0172	0.0251	0.0856	-0.0569	0.1864	0.0910	0.0000	0.0081	0.0072	0.3292**
7	-0.0140	0.0172	-0.0080	0.0078	0.0884	0.0003	-0.0084	0.0098	0.0359	0.1290**
8	-0.0168	0.0242	0.0516	-0.0434	0.1241	0.0210	-0.0024	0.0349	0.0337	0.2269**
9	-0.0269	0.0328	0.0164	-0.0229	0.1993	0.0080	-0.0036	0.0143	0.0824	0.2996**

*T = Traits
Note: The description of traits are given in Table 3
Residual = 0.3228 ** significant at 1% level
* Significant at 5% level

The earlier reports on genetic variability and correlation studies (Lad *et al.*, 1999 and Nagaraja *et al.*, 1999) with limited number of accessions revealed strong positive correlation with yield by pods per plant, pod length and grains per pod. By these studies it has been reported that the desirable plant type in horsegram is one with high number of primary branches, pod bearing nodes and pods. The present study involving 965 accessions also revealed the importance of these traits in selection of horsegram germplasm for further studies. Exploration of the wide range of variability in the germplasm would go a long way in increasing the production in this crop.

In addition to the above, during *rabi* 2000-2001 and 2001-2002, 46 accessions of horse gram collected under NATP-PB, were characterized for eight qualitative and nine quantitative characters. The maximum variability was found in yield per plant followed by number of pods per plant in both the years and the least variability was found to be in seeds per pod and days to maturity. The PCV % obtained with respect to the quantitative characters based on two years data are given in Table 6.

Table 6. Variability parameters in horsegram germplasm

	Range		Mean		SD		PCV (%)	
Characters	2000-2001	2001-2002	2000-2001	2001-2002	2000-2001	2001-2002	2000-2001	2001-2002
Number of Primary branches	3.00-13.00	4.00-11.70	8.95	8.15	3.07	2.17	34.3	26.57
Days to 50% flowering	62.00-80.00	55.00-97.00	73	71.93	5.96	10.24	8.2	14.24
Number of pods per plant	27.00-223.00	4.00-11.70	94.24	67.39	48.32	31.1	51.3	46.14
Pod length (cm)	3.70-6.10	55.00-97.00	5.16	4.61	1.6	0.32	31	6.84
Plant height (cm)	52.00-120.70	9.66-163.70	86.73	97.91	16.02	16.19	18.5	16.53
Number of seeds per pod	5.00-7.00	4.07-5043	6.28	6.18	0.59	0.38	9.4	6.2
Days to 80% maturity	108.00-128.00	71.00-155.33	120.32	121.54	8.88	13.67	7.4	11.25
Seed yield per plant (g)	0.70-23.00	5.33-7.00	7.04	5.58	4.2	2.98	59.7	53.37
100-seed weight (g)	2.60-8.20	94.00-152.00	3.26	3.16	0.3	0.33	9.8	10.56

During *rabi* 2003-2004, 197 accessions were subjected to a multilocation evaluation trial at this station and at National Network Research Project on Arid Legumes at Jodhpur. The report received from Jodhpur showed that the crop grown at Jodhpur did not flower due to photosensitivity except for 3 accessions. Nine qualitative and eight quantitative characters were recorded at Thrissur centre. The detail results obtained on 197 accessions in quantitative characters are given below:

S. No.	Character	Mean (g)	Range	PCV (%)	Mean of best check: Co-1
1	No. of pods per plant	31.6	10.0-84.3	46.7	32.4
2	Pod length (cm)	4.7	2.8-6.7	13.9	4.1
3	Plant height (cm)	73.3	38.7-136.0	23.7	79.8
4	No. of seeds per pod	5.5	3.0-7.33	13.8	5.3
5	Seed yield per plant (g)	2.7	0.4-7.5	55.3	3.1
6	100-seed weight (g)	3.2	2.1-4.5	14.6	3.9

The indigenous collections IC 145306 (Bihar), IC 068602 (Kerala) and IC 071777 (Tamil Nadu) were forward superior in terms of seed yield

Biochemical Analysis

In horsegram, a wild relative identified as a new species namely, *Macrotyloma sar-garhwalensis* (IC212722) collected from Pauri District, Uttaranchal was analysed for protein content at NBPGR, New Delhi. It revealed that seed protein content was 38.35 $\pm$ 1.35 % which is more than in cultivated check varieties of horsegram, such as SK 2001 (21.09%), VLG-1 (21.06%) and Raipur local (22.80%) and also higher than that of the 434 accessions of kulthi germplasm maintained at the NBPGR Regional Station, Akola (Negi *et al*., 2002).

Seed germination studies

Seed germination was studied in 50 accessions of horsegram. The results indicated that seed and its germination traits in the accessions taken for the study differed in range, mean, SD, SE and PCV significantly at 1% level with respect to seed length, breadth, thickness, germination percentage, germination vigour, plumule length and radicle length. Coefficient of variation ranged from 7.211 in seed thickness to 35.538 in radicle length. Higher values of PCV obtained in the case of radicle length (35.538), plumule length (31.826) and seedling vigour (22.667) indicated the maximum genetic variability present in these parameters with respect to the accessions studied (Table 7).

Table 7. Germination studies in horsegram

Character	Mean	Range	SD	SE	PCV (%)
Length (mm)	5.354	4.45 - 6.08	0.446	0.063	8.336
Breadth (mm)	3.496	2.55 - 4.36	0.436	0.062	12.474
Thickness (mm)	1.974	1.46 - 2.32	0.142	0.020	7.211
Length x Breadth x Thickness	37.329	21.77 - 56.05	8.152	1.153	21.839
Hundred seed weight (g)	2.995	1.44 - 4.04	0.431	0.061	14.400
Germination percentage	84.840	54.00 - 100.00	15.597	2.206	18.385
Germination vigour	8.130	3.00 - 9.00	1.843	0.261	22.667
Plumule length (cm)	4.215	1.04 - 6.33	1.341	0.190	31.826
Radicle length (cm)	4.236	1.32 - 7.37	1.505	0.213	35.538
Number of seed colours	4.92	2.0 - 9.0	1.496	0.212	30.409

Results on simple correlation studies between 8 characters namely, length, breadth, thickness, germination percentage, germination vigour, plumule length, radicle length and hundred seed weight, showed positive correlation between length and breadth, between germination percentage and germination vigour, between plumule length, radicle length and 100 seed weight, all at 1% level; it was also positive between seed length and 100-seed weight, between seed breadth and seed thickness and between 100-seed weight and seed thickness at 5% level. The results showed that almost all the seed parameters were correlated with seed germination indicating a very close relationship between seed health and the germination process as expected in the case of orthodox seeds. The extent of seed polymorphism noticed in horsegram by way of seed size and colour do not appear to be detrimental to seed viability as the germination % varied from 54 to 100.

Documentation

A catalogue has been published by NBPGR, New Delhi based on evaluation data generated on 11 economically important traits on 506 accessions evaluated at Delhi and 645 accessions at Akola during *kharif* 1984 and 1985. In addition to this, data on 275 accessions evaluated during 1988-1989 at Akola has been incorporated in that catalogue (Patel *et al.*, 1995).

Conservation

Long-term storage

Out of the 1040 accessions maintained by this station, 988 were sent to National Genebank at NBPGR, New Delhi for long-term storage, in which 551 were conserved at -20°C in the seed store. A total of 1493 indigenous and 12 exotic accessions have been conserved in the National Genebank so far.

Medium-term storage

In the medium-term storage facilities at NBPGR Regional Station at Akola and at Thrissur 1066 and 1040 accessions have been conserved respectively.

Germplasm Supply

A large number of accessions of horsegram has been supplied to several user agencies by this station so far for crop improvement and multilocation evaluation (MLE) as follows:

Year	Institute	No. of accessions
1980	College of Horticulture, Vellanikkara	50
1982	College of Agriculture, Trivandrum	121
1983	RARS, Kerala Agricultural University, Pattambi	14
1985	College of Horticulture, Vellanikkara	1
1985	University of Burdwan, West Bengal	19
1985	College of Agriculture, Trivandrum	197
1988	VPKAS, Almora, Uttar Pradesh	21
1988	College of Agriculture, Trivandrum	50
1989	University of Burdwan, West Bengal	50
1991	University of Burdwan, West Bengal	68
1991	College of Horticulture, Vellanikkara	1
1994	College of Horticulture, Vellanikkara	205
1994	College of Horticulture, Vellanikkara	13
1995	RARS, Kerala Agricultural University, Pattambi	100
1996	RRS, University of Agricultural Sciences, Tiptur	6
1996	RARS, Kerala Agricultural University, Pattambi	43
2002	ANGRAU, Hyderabad	10
2002	RRS, CSKHKV, Sirmour, Himachal Pradesh	70
2002	VPKAS, Almora, Uttar Pradesh	100
2002	Birsa Agricultural University, Ranchi, Jharkand	25
2003	RARS, Kerala Agricultural University, Pattambi	150
2005	University of Agricultural Sciences, Bangalore	435
2005	RARS, Kerala Agricultural University, Pattambi	11
2005	University of Agricultural Sciences, Bangalore	100 (MLE)
2005	RARS, Kerala Agricultural University, Pattambi	100 (MLE)
2005	Krishi Vidyapeeth, Rahuri	100 (MLE)
2005	Mahatama Phulke Krishi Vidyapeeth, Solapur	100 (MLE)

Germplasm utilization

Limited hybridization studies have been conducted in horsegram. Hybridization studies conducted between photosensitive and day neutral varieties with black and brown coloured seeds revealed that photoperiodic response is a qualitative trait, controlled by at least two genes, either in complementary or inhibitory gene action. In case of inheritance of seed colour, the black seed colour was observed to be dominant over brown. Two genes in polymeric gene action were found to control seed colour (Sreenivasan, 2003). Best utilization of germplasm for the development of improved varieties has been done in horsegram. Almost 99% varieties released for different states in India originated from local germplasm following their effective and specific evaluation. The varieties developed using germplasm in different states included BR5, BR 10 and Madhu for Bihar; HPK 2, HPK 4, HPK 5 and HPK 6 for Himachal Pradesh; PDM 1 and VZM 1 for Andhra Pradesh; K 82 and Birsa Kulthi for Jharkhand; S27, S 28, S 39 and S 1264 for Orissa; Co 1, 35-5-122 and 35-5-123 for Tamil Nadu; Hebbal Hurali 2, PHG 9 and KBH 1 for Karnataka; Maru Kulthi, KS 2, AK 21 and AK 42 for Rajasthan and VL Ghat 1 for Uttaranchal. Some of the improved varieties developed through single plant selection from the bulk collected included CO-1, No-35, 5-122, and 123 from Madras; Sel. 33, Sel. 34, Sel. 36 and Sel. 42 from Hyderabad; and black kulthi from Mysore. Some more promising varieties/selections developed in the last decade are 8-1-1-8, Belgaum 1-8-3, Bijapur 1-6-5, Hyderabad A-3-2-3; and Hebbal Hurali 1 and 2. Out of these, Hebbal Hurali 1 and 2 were developed by selections made from lines PLKU-32 and EC1460, from the material supplied by NBPGR, New Delhi (Kumar, 2005).

Multi-location evaluation

In recent years, much emphasis is placed on multi-locational evaluation of germplasm and the NBPGR is playing a leading role in this work in India. Under the National Network Research Project on Arid Legumes, NBPGR Regional Station, Thrissur was designated as the provider of 100 accessions of germplasm of horsegram for multilocation trial for 4 institutions every year. Last year 100 accessions were sent to 4 locations *viz.*, University of Agricultural Sciences, Gandhi Krishi Vignyan Kendra, Banglore; Regional Agricultural Resaearch Station, Pattambi, Kerala; Krishi Vidyapeeth, Rahuri, Maharashtra and Mahatama Phulke Krishi Vidyapeeth, Krashak Bhavan, Solapur, Maharashtra. This process will be continued and all the available accessions will be tested under multilocations.

Future Strategies in Collection and Conservation

- On the basis of gap analysis, the areas such as Bhilwara, Chattisgarh, Pratapgarh, Raipur and Pattambi have been identified as potential for germplasm collection with emphasis on local named landraces.
- Increased efforts are to be made to conserve the released varieties of various institutions for conservation on long-term basis.

References

Dewey, D.R and K.H. Lu 1959. A correlation and path coefficient analysis of components of crested wheat grass population. *Agron. J.* 51: 515-518.

Dobhal, V.K and J.C. Rana 1994a. Multivariate analysis in horsegram. *Legume Research* 17(3-4): 157-161.

Dobhal, V.K and J.C. Rana 1994b. Genetic analysis for yield and its components in horsegram (*Macrotyloma uniflorum*). *Legume Research* 17 (3-4): 179-182.

Gaur, R.D. and L.R. Dangwal. 1997. A new species of *Macrotyloma* from Garhwal Himalaya, U.P. India. *Bombay Natural Hist. Society* 94: 381-383.

Hanet, P. 2001. Mansfeld's *Encyclopedia of Agricultural and Horticultural crops* (Except Ornamentals), Springer, Berlin.

Jayan, P.K. and Maya C. Nair 2001. Studies on germplasm collections of Horse gram. *Indian J. Plant Genet. Resour.* 14 (1): 443-447.

Joshi, S.S., Chikkadevaiah and H.E. Shashidhas 1994. Classification of horsegram germplasm. *Crop Research* 7(3): 498-501.

Kumar, D. 2005. Status and direction of arid legumes research in India. *Indian Journal of Agric. Sci.* 75(7): 375-91.

Lad, D.B., S.A. Chavan and A.D. Dumbre 1999. Genetic variability and correlation studies in horsegram. *J. of Maharashtra Agri. Univ.* 23(3): 268-271.

Matthew, K.M. 1983. *The Flora of the Tamil Nadu Carnatic. Part One Polypetalae*. The Rapinat Herbarium, St. Joseph's College, Tiruchirapalli.

Nagaraja, N., S.D. Nehru and A. Manjunath 1999. Plant types for high yield in horsegram as evidenced by path coefficients and selection indices. *Karnataka J. of Agri. Sci.* 12(1-4): 32-37.

Nair, N.C. and A.N. Henry 1983. *Flora of Tamil Nadu, India Series I : Analysis, Vol. I.* Botanical Survey of India Southern circle, Coimbatore. pp.1 – 184.

Negi, K.S., S. Yadav, S. Mandal and D.C. Bhandari 2002. Registration of Wild Horsegram Germplasm (INGR No. 02007; IC212722). *Indian J. Plant Genet. Resour.* 15(3): 300-301.

Pandya, R., B. Ram, B. Vaid and L.K. Sharma 2003. Correlation and path analysis for yield in horsegram under two rainfall regimes. In: *Advances in Arid Legume*

Research. Eds. Henry, A., D. Kumar and N.B. Singh. Pp 123-125.

Patel, D.P., B.S. Dabas, R.S. Sapra and S. Mandal 1995. Evaluation of horsegram (*Macrotyloma uniflorum* (Lam.) germplasm. National Bureau of Plant Genetic Resources, New Delhi.

Patil, J.V., R.B. Deshmuch and S. Singh 1994. Selection of genotypes in horsegram based on genetic diversity. *J. Maharashtra Agric. Univ.* 19: 412-414.

Pullaiah, T. and E. Chennaiah 1997. *Flora of Andhra Pradesh (India). Volume 1. Rananculace-Alangiaceae*. Scientific Publishers (India), Jodhpur.

Purseglove, J.W. 1974. *Dolichos uniflorus*, In "Tropical Crops: Dicotyledons." London: Longman. pp. 263-264.

Savithriamma, D.L., Shambulingappa, K.G. and M.G. Rao. 1990. Evaluation of horsegram germplasm. *Current Research* 19(9): 150-153.

Sharma, R.K. 1995. Nature of variation and association among grain yield and yield components in horsegram. *Crop improvement* 22 (1): 73-76.

Sreenivasan, E. 2003. Hybridisation studies in horsegram. In: *Advances in Arid Legume Research*. Eds. Henry, A., D. Kumar and N.B. Singh. Pp 115-119.

Tripathi, A.K. 1999. Variability in horsegram. *Ann. of Agri. Res.* 20(3): 382-383.

3

Genetic Improvement

A. Henry, K.P. Viswanatha, S.K. Sharma and H.L. Thakur

Horsegram, a crop of great medicinal value, source of higher content of vegetable protein, is grown in large area of country under rainfed situations with poor fertility and less after care, has not been given due attention as far its genetic improvement is concerned. The crop is known to have many inbuilt defects and demerits like longer maturity, poor harvest index, photo and thermo-sensitive nature, susceptibility to some diseases, which warrant its genetic upliftment for augmenting productivity *per se*. The crops has however, been studied for few components of genetic improvement only. The information on some of these components, is still inconclusive and scattered in piece meals. An attempt has been made to compile such information which may be helpful in its genetic furtherance.

Genetics

Information on genetics of important traits is required for initiating breeding strategy, but too little reports are available on it, particularly on genetic aspects of grain yield and yield traits. Sen and Bhowal (1959) in a single intervarietal cross in *D.biflorus* reported that black seeds and purple plant colours were dominant over brown seeds and green pods and depended on a single gene or two closely linked genes. The earliness in flowering of black seeded parent was dominant over the late flowering of the brown seeded one. Genes for pigmentation and time of flowering were inherited independently. Kulkarni *et al.* (1978) studied inheritance of seed colour and purple eye on standard petal in horsegram in F_1 and F_2 of crosses between varieties EC-7460 (with a black mottled testa and without marking on the standard petal) and varieties Macharishore and Co-1 (both with a brown testa and a prominent purple eye on the standard petal). The results indicated that the presence of black mottling on the testa and of a purple eye on the standard petal were conditioned by single, independent, dominant genes designated as B^{11} and Pe, respectively. The interaction between two duplicate genes and inhibition of either of these two by an inhibitory gene were responsible for pigmentation on calyx. The segregation pattern of pigmentation of flower buds and flowers were digenic with different nature of interactions. The pigmentation were influenced by inhibitary action of other common genes. Almost absence of genetics of quantitative traits for their inheritance pattern and gene action, appear to have hampered efforts on yield enhancement of horsegram through manipulation of genes.

Polyploidy

Horsegram is an autogamous is nature, however some attempts to induce polyploids have been made. For instance, Sen and Vidya Bhusan (1959) attempted to induce polyploidy in 3 early and 4 late maturing varieties by colchicine treatment of the seeds and apical buds of one week old seedlings, seed treatment was however not successful. The treatment of apical buds resulted in development of polyploid sectors in young plant. Pure tetraploids were obtained from the same in the following generations. Diploid and tetraploid seeds germinated at about the same time and germination percentage in the tetraploids was equal to or slightly lower to that of the diploids. The tetraploids were marked by slow growing initially but eventually surpassed the diploids and produced fewer but thicker stem with larger flowers. Flowering was delayed in the tetraploids but was less marked in the late maturing varieties. The length of flowering period was increased by about 4-6 weeks. Tetraploids produced a large number of pods but these were smaller than in the diploids and contained fewer seeds. Since, pollen fertility was high (80-90%), it was suggested that reduced seed setting in tetraploids may have been due to zygotic sterility. A number of aneuploids also appeared in the progeny of selfed tetraploids.

Genetic Resource

A wealth of genetic variability is the prime requirement in any crop improvement programme. Greater the diversity in a crop, better are the chances of evolving desirable plant types. Genetic variability is the real measure of diversity concealed in a population, while the expressed variability or phenotypic variability is due to genotype, environment and their interaction. Thus, a detailed study on the extent of variability available for different characters associated with yield forms the initial step in any breeding programme including horsegram.

Nayar and Dhillon (2002) reported that NBPGR has 1543 accessions of horsegram out of which 910 lines have been characterized for 14 characters. Photo- insensitive and early maturing (90-100 days) genotypes with high yield potential have also been identified. The information generated from evaluation of these materials has been documented. The Tamil Nadu Agricultural University also maintains about 320 accessions at its Regional Research Station, Paiyur. Around 120 accessions are being maintained at GKVK, Campus of University of Agricultural Sciences, Bangalore. Besides, large germplasm is maintained at Dryfarm Research Station, Bhilwara and S.D. Agricultural University, S.K. Nagar.

Quantitative Studies

Genotype - Environment interaction studies

Stable performance of varieties under different environments is of considerable significance for any varietal improvement programme including horsegram. Information on genotye x environment interactions and stability of varieties were worked out by Deshmukh *et al.* (1986), Henry and Daulay (1988), Birari *et al.* (1990). Savithramma *et al.* (1996) evaluated promising genotypes of horsegram for stability parameter and found that genotype B-5-12-2 and BGM-2 were promising from the stability view point. Tah and Dasgupta (1997) carried out G x E interaction studied for protein content in two separate set of experiments and reported highly significant differences for seed protein (26.2-30.6%) content across the six environments. Significant genotypes x environment interactions were observed in breeding lines of certain cross combination of Kulthi (Tah and Dasgupta (1997). Venkateswarlu (2000) worked out stability parameters for twenty varieties and found that genotypes PHML-42 and Palem-2 were stable ones in different situations. Ram *et al.* (2002) evaluated variability parameters in 18 genotypes for grain yield and yield components in two environments (E1-400 mm and E2-615 mm rainfall) and significant variations were observed for days to flowering, days to maturity, plant height, branches/plant, pods/plant, seeds/pod, 100-seed weight, plant stand at maturity and grain yield.

Variability and correlation studies

The fundamental theorem of natural selection emphasized the importance of genetic variability. Greater the genetic variability upon which selection for fitness may act, the greater is the improvement in fitness.

Horsegram possesses wide range of genetic variability that can be exploited for genetic improvement of economic traits including high grain yield. Substantial variability for the traits like, plant height, number of branches, number of nodes, number of pods, days to flowering, days to maturity, pod length, pod yield and grain yield have been reported in horsegram by various workers (Ramakrishnan *et al.,* 1978, Ganeshaiah *et al.,* 1982, Suraiya *et al.,* 1988, Singh 1990, Balan *et al.,* 1992, Dobhal and Rana, 1994, Rao and Nanda, 1994, Sood *et al.* 1994, Savithramma *et al.* 1996, Nagaraja 1997, Samal and Senapathi, 1997 and Lad *et al.,* 1998. Sreekantaradhya *et al.,* 1975 obtained highest GCV values for number of nodes, followed by number of branches and number of pods. The values were however, moderate for seed yield and plant height and low for seeds per pod and pod length. Ganeshaiah (1980) studied 100 horsegram varieties, which showed significant variation in all the eight characters studied and observed that genotypic and phenotypic variations were greatest for number of secondary branches. Shivashankar *et al.* (1977) also studied hundred genotypes of horsegram and reported that genotypic variances were high for nodes per plant and number of secondary branches, moderate for pods per plant and days to flowering; and low for seeds per pod. They also suggested that variations were mostly due to genetic factors. Maximum variability for pod number, followed by plant height and node number was observed, which offered good scope for selection in crop improvement. Patil and Deshmukh (1982) reported considerable variability for plant height, days to flowering, maturity, number of pods, primary and secondary branches and seed yield/plant. However, range of variability was narrow for 100-grain weight and grains/ plant. Studies of Ganeshaiah *et al.* (1982) also revealed high estimates of GCV and PCV for number of branches, nodes and grain yield. Lowest values for GCV and PCV were however observed for seeds per pod, days to flowering, 100-seed weight and number of pods per plant. Patil and Deshmukh (1982) also reported that crude protein content in 50 genotypes varied from 17.94 to 28.88%. Similarly, Variability in protein content ranging from 19.9 to 25.8 and 14.8 to 26.9% was reported by, Kawde *et al.*, 1974 and Pore (1979), respectively. Borole (1984) observed low genotypic coefficient of variance for days to maturity in this crop. Kabir and Sen (1987) also reported significant genetic differences in the six yield component traits of horsegram. According to Suraiya *et al.* (1988), pods per plant, plant height, days to 50% flowering and days to maturity showed the highest, while 100-seed weight showed the lowest genotypic variance in horsegram. Sood *et al.* (1994) observed that the magnitude of phenotypic and genotypic coefficients of variation were similar for certain characters like seed yield, days to flowering , days to maturity and 100-seed weight, indicating little influence of environment in the expression of these characters. Savithramma *et al.* (1996) while evaluating 103 horsegram genotypes during kharif, rabi and summer seasons, observed that the differences between PCV and GCV were narrow in all the three seasons in case of days to flowering, days to maturity, seeds per pod, 100-seed weight, threshing percentage and protein percentage, indicating less influence of environment on expression of these traits and phenotypic values will be reliable for selection. Samal and Senapathi (1997), reported wide variation for plant height, pods/ plant, days to 50% flowering, number of branches/plant and yield/ plant but 100-seed weight showed minimum variance. Both GCV and PCV were low, for days to flowering, indicating less scope for its improvement through selection. Influence of environment as indicated by the difference between PCV and GCV, was very low in most cases, except for number of branches/plant. Lad *et al.*, (1998) reported wide range of genetic variability for yield and yield components. The phenotypic variance was higher than the genotypic variance for almost all the characters studied. The magnitude of genotypic variance was greater for almost all the characters except number of branches/plant, pod length, grains/pod and 100-grain weight. Nagaraja (1997) reported low estimates of PCV and GCV for days to flowering and days to maturity but high estimates in respect of primary branches. Low variability was also observed by many workers for certain traits like, pod length and number of seeds per pod by Sreekantaradhya *et al.* (1975), Balan *et al.* (1991), Savithramma *et al.* (1996) and Rao and Nanda (1994), days to flowering; and days to maturity and 100 seed weight by Samal and Senapathi (1997), Sood *et. al.* (1994), Singh (1990) Rao and Nanda (1994). Kabir and Sen (1987, 1989) reported significant genetic differences in 7 strains of *D. lablab*, and 8 strains of *D.biflorus* and 28 F_1 plant. The estimates on GCV and PCV were highest for pod yield per plant and heritability values were higher for all the characters. Similarly, Gupta and Samanta (1991) provided information on genetic variability and heritability in seven *Lablab purpureus* and one *D. biflorus* genotypes and their F_1 hybrids.

Correlation and Path studies

Estimates on genotypic correlation coefficients provide a measure of genotypic association between the characters. Correlation indicates that how improvement in one character will simultaneously cause change in other. High positive correlation was reported for grain yield with its components viz. pods/plant, seed size, pod length, plant height and number of branches/ plant (Aggarwal and Kang, 1976). Many workers have also reported high positive correlation of seed yield with seeds/ pod and nodes/plant (Shivshankar *et al.,* 1977); pod number, secondary branches and 100-grain weight (Patil and Deshmukh, 1983). However, days to flowering and maturity were negatively correlated with grain yield, seeds/pod and nodes/plant (Sivshanker *el al.,* 1977;

Ghorapade, 1985). On the contrary, Ghorpade (1985) observed that seed yield was negatively correlated with pods per plant and seed index was positively correlated with 50% flowering. He also classified 216 genotypes into 3 groups through Metroglyph analysis on the basis of 50% flowering and grain yield was positively correlated with days to first pod maturity, pods per plant and seeds per pod. A negative correlatin was found between 100-seed weight and certain other characters (Birari *et al.,* 1987). Dabas *et al.* (1990) indicated that seed yield and plot yield were positively correlated with each other and also with number of clusters on the main stem, on main branches and per plant, pods on branches and per plant. Strong correlation with yield was observed with pods/plant, pod length, grains/pod and dry weight of pods/plant (Lad *et. al.,* 1998). Tripathi, (1999) reported that early maturity, higher number of branch, number of pods and higher seed weight may be selected for higher yield. Patil and Deshmukh (1981) demonstrated that number of pods/ plant showed maximum direct effect on yield followed by number of secondary branches per plant. High positive direct as well as indirect effect of pods/plant on yield was also reported by Aggarwal and Kang (1976). Path analysis revealed that pods/plant and number of fruiting nodes/plant were major yield contributing characters (Kallesh, 1987). Lad *et al.,* (1999) reported that pods/ plant had significant high positive correlation with yield mainly because of substantial positive indirect effects via days to first flower, bud appearance, pod length, grains/ pod, test weight, number of compound leaves/ plant; days for overall maturity, dry weight of plant and grain yield. The significant positive correlation of leaf area with yield was due to the direct positive effect plus the positive indirect effect via plant height, number of branches/plant, pod length, grain/pod, test weight, days for maturity, dry weight of plants and number of compound leaves. Clusters per plant exhibited the greatest direct effect on grain yield followed by days to flowering and pods per plant (Dobhal and Rana, 1994). Sharma (1995) indicated the importance of pod length, number of pods per plant, number of grain per pod and 100-grain weight for grain yield. The grains/plant showed strong positive genotypic correlation with pods/plant, branches/plant and plant height. Path coefficient analysis revealed that days to 50% flowering and pods/plant had maximum direct effects on grain yield/plant. Moreover, pods/plant had indirect effects on yield via days to 50% flowering, plant height and number of branches/plant. Thus, pods/plant, plant height, days to 50% flowering may be considered in selection for increased grain yield (Chakrabarty and Singh, 2002) of horsegram.

Heritability and Genetic Advances

High estimates of heritability and high genetic advance were reported for grain yield/plant, 100-seed weight and number of pod/plants in horsegram (Sreekantradhya *et al.,* 1975, Agrawal and Kang, 1976 and Ram Krishnan *et al.*, 1977. In another study heritability estimates for test weight were maximum and higher magnitude of estimates for the same were noticed for maturity, leaf area, yield/plant, pods/plant, dry weight; whereas plant height and grains/pod had relatively lower magnitude of heritability (Borde, 1984). Shivashankar *et al.* (1977) indicated high genetic advance for number of secondary branches and nodes. High heritibality values were obtained for number of primary branches, secondary branches, primary branch length, 50% flowering, node number and 100-seed weight, whereas lower for plant height, seeds per pod, pods per plant and yield. Ganeshaiah (1980) reported the highest genotypic and phenotypic variations for number of secondary branches and the heritability values were high for days to flowering and maturity. Patil and Deshmukh (1982, 1983), during two years study reported high estimates of heritability and genetic advance for the characters, viz., seed yield, number of secondary branches, primary branches and pods per plant. Seed yield was positively associated with pods per plant, number of secondary branches and 100-seed weight. Birari *et al.* (1987) found high estimate of genetic advance for yield per hectare and seeds per pod, while high heritability was exhibited by yield per hectare, 100-seed weight and days to first pod maturity. Kallesh (1987) studied 10 characters in 3 F_2 populations of horsegram and indicated high estimate of heritability and genetic advance for primary branches, number of seeds per plant and pods per plant. Suraiya *et al.* (1988) reported high values for heritability and genetic advance for days to 50% flowering, days to maturity, plant height and branches per plant; while low heritability was exhibited by seed yield per plant and number of pods per plant. Singh (1990) indicated that pods number per plant was the most important yield component in horsegram. Rao and Nanda (1994) reported low estimates of heritability for all the characters. Dobhal and Rana (1994) reported the highest, PCV, GCV, heritability and GA for internode, clusters per plant and days to flowering. Samal and Senapati (1997) reported high heritability and genetic advance for pods per plant, seeds per pod and yield per plant. Senapati *et al.,* 1998 reported high heritability and moderate genetic advance estimates for seeds per pod; moderate heritability and high GA for branches per plant and moderate heritability and moderate GA for pods per plant. Nehru *et al.* (2001) observed high genetic advance for number of primary branches and biomass yield/plant, moderate for grain yield and lower for pod length, number of primary branches, seeds/ pod, nodes on main stem and pods on main stem. Venkateswrulu (2000) observed high estimates of genetic advance and heritablity for pods/plant and clusters/plant. High estimates of heritability (broad sense) were observed for 50% flowering, grain yield/plant and pods/plant. Days to maturity and 100-seed weight showed high heritability with moderate genetic advance (Chakrabarty and Singh, 2002).

Genetic Divergence

Ramakrishna *et al.* (1979) studied genetic divergence in 8 yield components of horsegram and observed that chief contributors to total divergence in horsegram were 100-seed weight and dry weight of nodular tissue. No association was found between geographical diversity and genetic diversity. Ganeshaiah (1980) reported that the characters contributing most to the diversity of material were mainly associated with plant maturity but there was no clear cut association between genetic and geographical diversity. Balan *et al.* (1992) analyzed grain yield and 8 component traits for genetic divergence and genotypes were grouped into 9 clusters, but none of which were related to their geographic origin. Pod weight was found contributing the highest D^2 values. Dobhal and Rana (1994) studied genetic divergence in 23 genotypes of horsegram in respect of 9 characters; and varieties were grouped in to seven clusters and concluded that genetic diversity had less relevance to eco-geographical distribution of genotypes. Canonical analyses indicated that pods per plant, days to flowering, yield per plant, pod length and branches per plant contributed the most to the divergence. Patil *et al.* (1994) based on D^2 grouped 50 genotypes of horsegram into 13 clusters. Days to flowering and maturity, plant height, number of primary and secondary branches and pods per plant were highly divergent among the characters.

Mutation Breeding

The study on DES for studying LD 50 of horsegram (var. Hebbal Hurali-1) indicated that the same was 0.03 per cent (Kulkarni *et al.*, 1978). The effect of gamma rays and ethyl methane sulphonate on horsegram revealed that the frequency of chlorophyll mutants increased with increasing EMS concentration up to 1.2% and decreased thereafter in M_2 generation. But seed treated with gamma rays revealed that there was an increase in mutation frequency with increasing doses from 5 to 15 kR and decreased thereafter, in M_2 generation. The frequency of macro-mutations decreased with increase in EMS concentration. A wider spectrum and greater frequency of macro-mutations was induced by EMS than by gamma rays. The gamma ray treatment resulted in only two types of macro-mutations, viz. dwarfness and testa colour (Kulkarni and Shivashankar, 1978). The effect of gamma radiation (1.25-15 kR) in *Dolichos, Vigna* and *Phaseolus* species was studied by Sinha and Himansu (1984) and they reported that in case of horsegram, chlorophyll contents increased in all the doses. However, in other legumes it decreased at lower doses and increased at higher doses. Rudraswamy (1987) reported induction of chlorophyll mutants, viz., Albina, Xantha and Viridis in 3 varieties of horsegram. Gamma rays, in general, found to induce a wider spectrum of chlorophyll mutants than did EMS. He obtained mutants for the characteristics, viz., leaf colour, testa colour, dwarfism, earliness and photo insensitivity. Suresh *et al.,* (1991) studied induced variability for grain yield and dry matter production in M_3 progeny of two varieties (Kurung Kollu and CO-1) irradiated with 10-80 kR gamma rays. Results revealed that in case of var. Kurung Kollu, mean increased and shifted positively in all 648 M_3 progeny populations, whereas, only in 2 population (out of 603 M_3 progenies) of var. CO-1, it increased. Samal and Senapati (1997) irradiated the seeds of local horsegram lines with 20,40 and 60 kR gamma rays during 1991 and 15 selected mutants were evaluated with 3 standard checks in M_4, their results further indicated that KHG 60-5 was the highest yielder followed by KHG 40-8 and KHG 20-3 mutants.

Bakle and Pochi (1991) reported the effect of mutagen agents i.e. gamma rays (5-80 kr), ethyl methyl sulphonate (EMS) and diethyl sulphonate (DES). The lower doses of gamma rays showed stimulatory effects on germination whereas, all the concentrations of EMS resulted in decreased germination, except in case of pre-soaked EMS treatment (0.6%). The increased doses and durations of mutagenic agents resulted in various abnormalities in plant height, leaf numbers, leaf size and shape and induction of chlorophyll chimeras. Lower doses (5-10 kR) and concentration (0.6-0.9%) resulted in increased number of pods, seed yield and protein content. The chromosomal abnormalities also increased with increased doses. Effects of gamma-rays and EMS on seed fertility and chlorophyll in cv. HPK 4, HH 2 and BGM 1 were studied at Banglore. Mutagens resulted in severe inhibition of growth and reduced accumulation of food in developing seeds, which may be due to accumulation of chromosomal aberrations, cytoplasmic mutations and phenological effects (Vishwanath *et al.,* 2002). Gamma rays were most effective in inducing chlorophyll mutations in M_1 and M_2 generations and Albino and Xantha type of mutations were more frequent.

Varietal Selection

Genetic improvement work in horsegram was mostly done during late fourty's and fifty's under ICAR sponsored pulse improvement schemes. For instance in Bombay, three selections were found superior to the local check at Dharwad. In Madras state, culture No. 35, 5-122 and 123 were superior to local bulks. In Orissa, selections 27, 28, 39 and 164 were of promise. In Hyderabad, selections 33, 34, 36 and 42 were superior to check. In Mysore state, a black Kulthi was found promising. The other promising selections were 8-1-1-8, Belgaum 1-8-3, Bijapur 1-6-5 and Hyderabad A-3-2-3. A renewed attempt at the improvement, started with the start of All India Coordinated Pulse Improvement Project and an ad hoc project in Karnataka in late sixty's and early seventy's.

Recently some centres have been identified under All India Coordinated Pulses Improvement Project to intensify multidisciplinary improvement work (Madurai/ Coimbatore in Tamil Nadu, Bangalore in Karnataka and Hyderabad in Andhra Pradesh). Nezamu-ddian (1970) reported a variety Co-1 developed by single plant selection from Mudukulathur in Ramanathapura district of the Madras state, which was superior in yield over local bulk by about 25-50 per cent. Sivashankar *et. al.* (1976) reported the evolution of two improved varieties for Karnataka state, *viz.* Hebbal Hurali 1 and Hebbal Hurali 2 by selection from lines PLKV-32 and EC 1460, respectively. The main achievement observed in these lines was in terms of reduction in maturity period (from 120 to 90-100 days), higher yield potential (18 to 20 q/ha) and increase in protein content from 19.5 to 21.5 per cent. Being photo-insensitive, these lines could be sown all the year round. The other improved varieties developed mainly by single plant selection from local bulk are as follows:

- Bihar : BR-5, BR-10 and Madhu
- Himachal Pradesh : HPK-2, HPK-4, HPK-5 and HPK-6.
- Andhra Pradesh : PDM-1, VZM-1.
- Rajasthan : Maru Kulthi, AK-21, AK-42.
- Orissa : DS-2-1-1, DS-2-2, OGH-11, OGH-60-1
- Maharashtra : Dapoli local, KBR-1, Local red, SINA, Man
- Tamil Nadu : Paiyur-1, Co-1, DPI-1236, Kurunkollu

In addition to it a few other varieties have been developed in different states viz. BR 5 and BR 10 (Bihar), S 27, S 28, S 39, S 1264 (Orissa), 35-5-122, 35-5-123 (Tamil nadu), DS 1-2, Birsa Kulthi, Maru Kulthi/1, and K 82 (Jharkhand). All these varieties have been developed through direct selection of local germplasm. For the first time the genotypes adapted in hilly regions of Himachal Pradesh were subject to selections and testing over a wide range of environments. These genotypes are also being used in hybridization programmes as a diverse gene pool never before introgressed with peninsular types.

Varietal Development

Several varieties of horse gram differing in seed colour and maturity are under cultivation. The seeds may be brown, light red, grey, black or mottled. The important varieties grown in the country and their distinguished features are given in Table 1.

Table 1. Varietal status of horsegram in India.

Variety	Maturity (Days)	Yield (kg/ha)	Special characteristics	States of adaptation
CO-1	110-115	600-700	Suited for rainfed areas. It is a selection from Munduka Lathur area. Plants are trailing with brown seeds. It was released in 1953 for Tamil Nadu.	Tamil Nadu
Paiyur-1	110-115	650-750	Suited for rainfed areas.	Tamil Nadu
Paiyur-2	105-110	700-900	High production, suited for sesame-groundnut-ginger-horsegram crop sequence in rainfed region of T.N.	Tamil Nadu
Hebbel Hurali 1	90-100	1800-2000	Selection from PLKU 32 and, it is photo-insensitive variety.	Karnatka
Hebbel Hurali 2	90-100	1800-2000	Selection from EC 1. It is photo insensitive variety.	Karnatka

Continue....

Variety	Maturity (Days)	Yield (kg/ha)	Special characteristics	States of adaptation
HPK-2	120-130	900-1200	Selection from local germplasm of Himachal Pradesh. It is a semi spreading variety with light yellow seed colour and seed weight 30-50 gm/1000 seed. It is selection from local germ plasm.	Himachal Pradesh
HPK- 4 (Baiju)	120-130	800-1000	It is almost the same as HPK 2, but seed is dark grey in colour and mottled in shape.	Himachal Pradesh,
Madhu	105-110	850-1000	Selection from local germplasm. It has bushy habit, seed colour is dark cream with red spots, grains medium bold.	Bihar
PDM -1	100-105	1400-1600	It is a local selection from Bhimali Patnam. Suitable for kharif sowing. Plant is spreading, flowers are yellow with light purple wash, seeds are buff coloured.	Andhra Pradesh
VZMI	100-105	1200-1500	It is a local selection from Vizia Nagaram. Plants are tall, semi erect, flowers are light yellow with purple wash. Seeds are black, suitable for kharif planting.	Andhra Pradesh
Birsa Kulthi-1	92-97	800-1000	A selection from local germplasm, released in 1985 for Bihar state, slightly trailing growth habit. Flowers light yellow with pink spot in the centre of petal, Seeds creamy, moderately resistant to *Macrophomina* leaf blight.	Bihar

Continue....

Variety	Maturity (Days)	Yield (kg/ha)	Special characteristics	States of adaptation
V.L. Gahat-1 (VLG-1)	125-130 (Hillls)	800-1000	A selection from local material collected from U.P. hills, released in 1983 for Uttar Pradesh hills. Plants semi-erect with spreading branches. Foliage light green and thin. Seeds medium in size, shining brownish yellow. Tolerant to blight.	Uttranchal
Maru Kulthi	90-95	800-900	Released in 1989 for all India. It is selection for local material, semi-spreading type, seeds are light brown.	Rajasthan
KBH-1 (BGM1)	100-110	900-1000	A selection from Bailhongal local, released in 1990 for the state of Karnataka. bushy, gives out tendril, photosensitive, tolerant to YMV, seed colour light brown.	Karnatka
PHG-9	100-105	1000-1200	A selection from Palampur local released in 1997 by CVRC for South India and in 1998 by SVRC (Karnataka) for Southern Karnataka	Karnatka
AK-21	85-90	800-1000	An early high yielding variety, suited for northern India	Rajasthan
AK-42	80-85	850-1100	Improved over PHG-9 and AK-21 early maturing variety released in 2005	Rajasthan

Tissue culture

Differentiation in callus and shoot buds were achieved in horsegram, only from cotyledon derived callus on MS medium, supplemented with B5 vitamins with different concentrations of NAA and BAP. Rooting was observed on MS medium containing 1.5 mg IBA/litre. Varisaimohamad, *et al.*, 1998 was successful in inducing multiple shoots from tip explants in *in vitro* grown seedlings on MS medium supplemented with ADS and BAP individually with different combinations and concentrations. Sinha and Das (1986) worked out anther derived callus of horsegram, its protoplast culture and morphogenic potential in var. BR-10. When anthers were cultured on SS-A8 callus induction medium, maximum response was obtained from anthers containing PMCs. This callus produced globular somatic embryoid on medium containing BA and GA_3. Same developed further on medium containing any of several auxins or auxins/ cytokinin combinations. Plantlets were recovered on SS-A8 medium supplemented with coconut milk. Mohammed and Jayabalan (1996) established, calluses of var. CO^{-1} and Bolgam-1 from explants of epicotyl, hypocotyls, cotyledons and young + cab on MS medium supplemented with auxins and cytokinins. Thus, tissue culture/ biotechnological efforts appear to be in quite initial stage and the same require momentum.

Future Strategies

1. Information on genetics of yield and yield components are meager, hence, are required for systematic breeding of horsegram for yield enhancement.
2. The crops requires early maturity so as to avoid yield losses due to terminal drought.
3. Efforts are required for breeding horsegram for photo and thermo-insensitity for increasing acreage in non-traditional areas and year round planting.

References

Aggarwal, V.D. and Kang, M.S. 1976. *Trop. Agri.* 53: 335-400.

Bakle, V.L. and Pochi, D.V., 1991. Mutational studies in *Dolichos bifloris* pp. 249-253 In: *Genetic Improvement of Pulse Crops* Vol. 2 (Eds. Jafar Nizam, Irfan. A. Khan and S.A. Frook).

Balan, A., P. Ramaswamy and Sivasamy, N. 1992. Genetic divergence in horsegram. *Indian Jour. Pulses Res.* 5:78-81.

Balan, A., Ramasamy, P. and Sivaswamy, N., 1991. Correlation and path Co-efficient analysis in horsegram, (*Macrotyloma uniflorus* (L.) Verde) *Orissa J. Agric. Sci.,* 4: 6-10.

Balan, A., V.D. Guruswamy, Raja P. Vivekananda 1991. Direct selection parameters in horsegram. *Legume Research* 14: 103-104.

Birari, S.P., N.V. Deorukhakar and Jamdagni, B.M. 1987. Genetic parameters and character association in horsegram. *Indian J. Agric. Res.* 21: 55-58.

Birari, S.P., V.V. Bhavsar and Jamadagni, B.M. 1990. Phenotypic stability for grain yield in horsegram. *Indian J. Agric. Sci.* 60: 683-684.

Borole, D.S.,1984. Genetic variability and correlation studies in horse gram. *M.Sc. Agri. Thesis*, Konkan Krishi Vidyapeeth, Dapoli, India.

Chakrabarty, M and Singh, D.N., 2002. Heritability and path coefficient analysis in horse gram. Paper presented in *National symposium on Arid Legumes for food nutrition security and promotion of Trade*, held at HAU, Hissar from May 15-16, 2002. pp. 54-55.

Dabas, B.S., T.A. Thomas and Sapra, R.L. 1990. Factor analysis in kulthi. *Intern. Jour. Trop. Agri.* 8: 14-18.

Deshmukh, R.B., A.H. Sonone, S.B. Patil and Kotte, T.B. 1986. Stability for grain yield in horsegram. *Curr. Reporter,* Mahatma Phule Agril. Univ. 2: 49-51.

Dobhal, V.K. and Rana, J.C. 1994. Maru Khethi variety analysis horsegram. *Legume Res.* 17: 157-161.

Ganeshaiah, K.N. Shivashankar.G. and Hiremath, S.R. 1982, Genetic variability for yield and other characters in the germplasm of horsegram. *Mysore J. Agril . Sci.,* 16(4): 400-403.

Ganeshaiah, K.N. 1980. Multivariate analysis for seed yield and its contributory characters in horsegram. *Mysore Jour. Agri. Sci.* 14: 125.

Ghorpade, D.S. 1885. Metroglyph analysis of Morphological variation in horsegram. *Curr. Res. Reporter* 1: 56-60.

Gupta, S.K. and Samanta, S.K. 1991. Genetic variability for green pod yield and other characters in Lablab bean. *Agril. Sci. Digest.* 11: 93-99.

Henry, A. and Daulay, H.S. 1988. Genotype x environment interactions for seed yield in horsegram. *Indian J. Agric. Sci.* 58: 794-795.

Kabir, J. and Sen, S. 1987. Studies on genetic variability and heritability in *Dolichos* bean. *Ann. Agril. Res.* 8: 141-144.

Kabir, J. and Sen, S. 1989. Correlation and path analysis in *Lab lab niger Medic*, *Dolichos uniflorus* Lam and their F_1. *Trop. Agri.* 66: 281-283.

Kabir, J. and Sen, S.K. 1987. Diallel analysis for combining ability in Dolichos bean (Lab lab Niger Medik and *Dolichus biflorous* Lamm.). *Trop. Agri.* 67 (2): 123-126.

Kallesh, H.G. 1987. Path analysis and selection indices in three F_2 populations of horsegram. *Mysore J. Agric. Sci.* 20: 335.

Kawde, R.S., Thakur, K.K. and Kadhe, N.N. 1974. *Indian J. Nutrit. diete.*11:83-85

Kulkarni, R.N. Shivshankar, G. and Mahishi, D.M. 1977. *Curr. Sci.* 46 :28

Kulkarni, R.N. and Shivshankar, G. 1978. Mutagenic effect of gamma rays and Ethyl Methane Sulphonate in horsegram. *Genetic Agraria* 32: 63-67.

Kulkarni, R.N., D.M. Mahishi, S.R. Hiremath and Chikkadevaiah 1978. preliminary studies on mutagenic effect of dimethyl sulphate on *Dolichos biflorus* L. *Curr. Res.* Bangalore 6: 101-102.

Lad, D.B., Chavan, S.A and Dumbre, A.D. 1998. *J. Maharashtra Agri. Univ.* 23: 268-271.

Lad, D.B., Chavan, S.A and Dumbre, A.D. 1999. *J. Maharashtra Agri. Univ.* 24 (1): 102-103.

Mohamed, S.V. and Jayabalan, N. 1996. A protocol for horsegram callus induction. *Israel J. Pl Sci.* 44: 143-145.

Nagaraj, N. 1997. Genetic variability and association studies involving seed yield, agro-morphological and seed characters in horsegram *(Macrotyloma uniflorum* Lam. Verdc.). *M. Sc. (Ag.) Thesis* UAS, Bangalore, India.

Nayar, E.R. and Dhillon, B.S. 2002. Genetic Resources in arid legumes- some important issues, *Ann. Report* pp 5-7.

Nehru, S.D., Majunath, A. and Nagaraja, N. 2001. Identification of traits to be selected in breeding yield in horsegram (*M. uniforum* L.), *Karnataka, J. Agri. Sci.,* 13(1): 11-15.

Nezamuddin, S. 1970. Miscellaneous pulse crops in India. (Ed. P. Kachroo), Publisher I.C.A.R., New Delhi, p. 306.

Patil, J.V. and Deshmukh, R.B. 1981. *Legume Res.* 4 : 62-64.

Patil, J.V. and Deshmukh, R.B. 1982. Genetic variability in horsegram. *Jour. Maharashtra Agril. Univ.* 7: 139-140.

Patil, J.V. and Deshmukh, R.B. 1983. Association analysis in horsegram. *Jour. Maharashtra Agril. Univ.* 8: 120-121.

Patil, J.V., R.B. Deshmukh and Singh, S. 1994. Selection of genotypes in horsegram based on genetic diversity. *J. Maharashtra Agril. Univ.* 19: 412-414.

Pore, M.S. 1979. *Indian J. Agri. Sci.* 49 : 532-534.

Ram, B., Pandya, R., Sharma L.K. and Vaid B. 2002. Hybridization studies in Horsegram. Paper presented at *National symposium on Arid Legumes for food nutrition security and promotion of Trade*, held at HAU, Hissar from May 15-16, 2002. pp 53-54.

Ramakrishna, A., Marappan, P.V. and Sivswamy, N. 1979. Genetic divergence in horsegram. *Indian J. Agric. Sci.* 49: 719-723.

Ramakrishnan, A., Marappan, P.V. and Sivswamy, N. 1978. Genetic diveeyence in horsegram. *Ind. J. Agric. Sci.,* 49 (9): 719-723.

Ramkrishnan, P., Muthu krishnan, C., Ranganathan, R. and Sivasamy, N. 1977. *Madras Agri J.* 65:281-283.

Rao, S.S. and Nanda, H.C. 1994. Genetic variability and correlation studies for grain yield and its components in horsegram. *Ind. J. Pulses Res.* 7: 59-61.

Rudraswamy, P. 1987. Induced mutagenesis in horsegram using gamma rays and E.M.S. *Mysore J. Agric. Sci.* 21: 95.

Samal, K.M. and Senapati, N. 1997. Genetic variability performance correlation and coheritability of horsegram mutants lines. *Legumes Res.*. 20: 207-211.

Savilkaramma, D.L., A.M. Rao, S. Ramesh, T.E. Nagaraja and Madhu Sudan, K. 1996. Stability analysis in horsegram. *Crop Imp.* 23: 111-114.

Savithramma, D.L. Nagaraj, T.E. Viswanatha, K.P. and Shambulingappa, K.G. 1996. Studies on genetic variability in horsegram. *Ind. J. Pulses Re.* 9(2): 127-131.

Sen, N.K. and Bhowal, G.J. 1959. *Indian J. Genet.* 19: 228-233.

Sen, N.K. and Vidya bhushan, R.V. 1959. *Sci. Cult.* 24 :523-524

Senapati, N., K.M. Samal and Dhal, A. 1998. Genetic variability and performance of promising horsegram lines under the north central plateau, agro-climatic zone of Orissa. *ACIAR Food Legume Newsl.* 27: 9-10.

Sharma, R.K. 1995. Nature of variation and association among grain yield and yield components in horsegram. *Crop Imp.* 22: 73-76.

Shivshankar, G., Chikkadevaiah and Hiremath, S.R. 1977. Germplasm of fieldbean. *Indian J. Genet.* 37: 353-371.

Singh, D.N. 1990. Genetic variability, correlation and path analysis for grain yield and its components in horsegram. *Indian J. Pulses Res.* 3(1): 25-30.

Sinha, R.R. and Das, K. 1986. Anther derived callus of *Dolichos biflorus* L. its protoplast culture and their morphogenic potential. *Curr. Sci.* 55: 447-452.

Sinha, S.N. and Himansu, R.S. 1984. Effect of gamma irradiation on chlorophyll metabolism in *Dolichos, Vigna* and *Phaseolus* species. *Cytologia* 49: 279-287.

Sood, B.C., Gartan, S.L. and Kalia, N.R. 1994. Variability, correlation and path studies in horsegram. *Indian J. Res.,* 7(1): 68-6.

Sreekantaradhya, R., Shetty B.I., Chandrappa, H.M. and Shivshankar, G. 1975. *Mysore J. Agri Sci* 9: 391-363.

Suraiya, D., S.T. Mercy, D. Chandramony and Nayar, N.K. 1988. Genetic variability of yield and its components in horsegram. *Indian J. Agric. Sci.* 58: 798-799.

Suresh, M., A. Narayanan, P. Vaidyanathan and Rangaswamy, S.R.S. 1991. Induced variability for grain yield and dry matter production in horsegram. *Mutation Breed. Newsl. No.* 37, 6-7.

Tah, J. and Dasgupta, P. 1997. Genetic organization of the minor pulses kulthi under micro environments of a location. *Indian Agricult.* 41: 233-248.

Tripathi, A.B. 1999. Availability in horsegram. *Annals Agri. Res.*, 20: 382-383.

Varisaimohamed, S., Jawahar, M. and Jayavalan, N. 1998, Effect of Ads, BAP and IBA on plant regeneration from *Macrotyloma uniform* (Lam.) Verdc. *Phytomorphol.,* 48: 61-65.

Venkateshwarlu, O. 2000. Phenotypic stability in horsegram (*Macrotyloma uniflourm* (L.) Verdc.) *Ann. Agri. Res.*, 21(1): 138-139.

Vishwanath, K.P., Rudraswani, P. and Rudresh, N.S. 2002. Mutagenic effect of gamma rays and ethyl methyl sulphonate in horse gram *Nat. Symp. Arid Legumes for food nutrition, security and promotion of trade at Hisar,* pp. 52-53.

4

Crop Production

A. Henry, S.N. Sodani and S.B.S. Tikka

Horsegram, a drought hardy crop is widely grown in India from 300-500 mm rainfall situations in 0-3^{0}N altitude at a temperature range of 20-35^{0}C. It is adapted to a wide range of soils like, deep red, loams, black cotton soils, clayey, paddy soils, sandy and shallow soils, stony and gravelly upland soils, rough jungles etc., as a sole crop and in a number of cropping systems. The crop is grown in kharif and rabi seasons also (July to October). Thus, it is desired that package of practices and various production technologies are summarized keeping in view its short day and photo-sensitive nature. An improved variety backed up by region specific production technologies would help increase production of this crop having marginal status and facing many production constraints. Hence, an attempt has been made to review different components of production technologies of horsegram concerning various growing regions.

Climate and Soil Requirement

Horsegram is grown as a rainfed crop in poor soils in which most other crops may not perform well. In north western India, it is grown with the onset of monsoon on poor marginal and sub-marginal lands. In Orissa, it is grown as a *kharif* crop in shallow red soils, gravelly soils; and hard upland soils of Kalahandi, Sambalpur and Cuttack and as a *rabi* (post-rainy season) crop other districts of Orissa representing poor sandy soils. In West Bengal, it is grown on acid lateritic soils (pH 5.5). In Karnataka, it is grown as a sole crop in poor red sandy soils as a late *kharif* season crop, relay crop in fingermillet, when the fingermillet crop is poor, and in double cropping system after *kharif* groundnut or sesamum and also when main crops like finger millet, sorghum and pearl millet are different be sown due to delayed monsoon. It is also extensively cultivated as an intercrop, in assured rainfall tracts of Karnataka comprising eastern belts lying immediately to the east of the western Ghats.

Efficient Cropping Systems

Intercropping systems with companion crops

Horsegram has been tried in a number of intercropping systems, like millets, oilseeds, grasses and trees at various centers in India. Research results on some of the same are as follows.

Oilseeds: It has been observed that horsegram can profitably be cultivated as intercrop without affecting the total income per unit area with hybrid sorghum and groundnut in a system of 2 rows of hybrid sorghum+2 rows of groundnut and intercropping horsegram when the groundnut crop is in flowering stage (Kambar and Kadappa, 1983). In Madhya Pradesh at Ambikapur, horsegam (var. Kulthi Birsa) gave the most promising performance in intercropping with niger over other legumes. The highest economic returns were obtained by planting one row of horsegram between two rows of niger at 40 cm row spacing (Bajpai *et al.,* 1990). In another study, niger when intercropped with horsegram among others in 2:1 row ratio, gave seed yield of 107-131 kg/ha as compared with 188 kg/ha in pure stand. The intercropped horsegram gave seed yield of 351 kg/ha which was the highest among all other intercrops tried viz., sesame, soybean and groundnut (Wali *et al.*, 1989). Intercropping of castor (var. VI-1) was studied in Maharshtra with cowpea, horsegram and clusterbean during 1984-88.

Millets: Three cropping systems viz., fingermillet alone or intercropping with horsegram or soybean were studied in Karnataka. The yield of fingermillet was highest (28 q/ha) when grown alone than while intercropped with horsegram (18 q/ha). Dry weight of weeds in these systems remained unaffected (Patil *et al.*, 1987). In Madhya Pradesh, intercropping of cowpea and horsegram with sorghum for fodder indicated better performance of cowpea over horsegram (Mishra *et al.*, 1997).

Maize: Horsegram alone was more profitable in the Alfisols of Bangalore than intercropped with maize. This is because maize fodder component of the intercropping system showed little sustainability under this agro-climatic conditions (Singh *et al.*, 1990).

Amaranth: Amaranth-horsegram intercropping system (1:1) produced the highest amaranth equivalent yield of 4.05 t/ha amongst the various legumes tried in row ratios of 1:1, 1:2 or 2:2 in Rani-chauri, U.P. (Arya *et al.*, 1996).

Grasses: Marvel grass (*D. annulatum*) when intercropped with horsegram, kidney beans (*P. americanum*) gave yield of 65.4 to 78.6 q/ha. The intercrops gave an average yield of 2.3-4.3 q/ha grain when grown on shallow soils of Solapur (Jadhav and Rathod, 1990).

Trees: In Bihar, intercropping studies with 12 tree species and 3 food crops was carried out during 1987. Soybean gave the highest average yield (6.31 q/ha) over all tree species, followed by sesame (5.67 q/ha), horsegram (4.57 q/ha) and niger (3.02 q/ha). Maximum yield (5 q/ha) was in combination with *Shorea robusta, Alibizzia procera* and *Azardirachta indica* (Jha *et al.*, 1989). In Karnataka, horsegram was grown between *Acacia nilotica* or *Azadirachta indica* rows spaced 5, 10 or 15 m apart with spacings of 1.2, 1.8 or 2.4 m. The horsegram yield decreased in agroforestry system. However, seed yield was the highest in trees spaced in 15 m rows (Basavaraju and Rao, 1997).

Allelopathy effects of trees on horsegram: In Anantpur (A.P.), allelopathic effects of *Digera muricata* (L.) Mart extracts (stem, leaf, inflorescence and root) on the germination and seedling growth of horsegram was investigated. Results indicated that seedling growth, length and dry weight of root and shoot of horsegram reduced hence, their combination was not compatible. Allelopathic effects of leachates and extracts of *Pinus roxburghii* were tested on four legumes including horsegram, which are generally cultivated around *chir pine* forests in terraced fields. It was found that the use of field soil (near the forest) increased the germination and growth of horsegram and black gram but reduced the same in soybean. Root extracts reduced the germination and growth of all legumes, however, shoot growth was increased in horsegram. Decaying tillers also reduced the germination and growth in most of the cases. Growing of horsegram and black gram near *chir pine* forest is recommended for better production (Singh and Verma, 1988). Allelopathic effect of aqueous leaf and bark extracts of some agroforestry trees on germination and radicle extension of horsegram among other legumes were tested. It was found that foliar and bark extracts of *Adina cordifolia* and *Grewia oppositifolia* inhibited the germination of horsegram and soybean while it did not affect other legumes tested. However, radicle extension of all the legumes were reduced by all the tree crops except *C. australis* (Bhatt *et al.*, 1993).

Double /sequence / rotation cropping systems: Fingermillet grown after cowpea and horsegram as fodder (10.2 q/ha) or green manure crop (11.5 q/ha) produced significantly higher yield over sole fingermillet (7.0 q/ha) in Karnatka. However, by growing a legume before fingermillet, the optimum date of sowing of fingermillet may be missed, which might result in reduced yield in a sequence cropping system (Singh and Venkateswarlu, 1985). In cropping sequence on uplands of Orissa, horsegram is the only crop can be taken after upland paddy, maize, sorghum and ragi (Acharya, 1983). Horsegram grown as a sequence during the dry season (after harvest of wet season crop) using residual soil moisture and nutrients appears to utilize the residual ^{15}N better when followed by the non-legumes compared to when it was followed by legume (Jena and Mishra, 1990). In another study, it was indicated that there was no significant difference in horsegram yield grown in the winter season following kharif rice, maize, sorghum or fingermillet. Net returns were higher when horsegram followed maize, sorghum or fingermillet (Nayak, 1993). Rotation experiments of various crops with sesamum were worked out in Tamil Nadu from 1962-65. Sesame grown after sesame, among other crops after horsegram gave average seed yield of 113 kg/ha. Sesame had no adverse effect on yield of any rotated crop (Chandrasekaran *et al.,* 1974). Studies on cropping system was worked out for 6 seasons during 1973-78 with native short grass vegetation in Sri Lanka. The pioneer crop sorghum cv. 1941 followed by *C.juncea, E.coracana* or *M.uniflorum* or maize composite followed by *E. coracana, M. uniflorum* or 2 season of lemon grass (*Cymbopogan citralus*) were grown before establishment of improved pasture and a control (without a pioneer crop was used). The results revealed that maize – horsegram combination gave the highest economic returns (Tharmarajah and Herath, 1988).

Efficient Soil and Moisture Management Techniques

On black cotton soils and mixed red and black soils, a system of broadbed-and furrows graded at slopes of 0.4 to 0.6 per cent has been successful for efficient soil and moisture conservation. The width of broadbed is fixed depending on the field size. In general, 3 to 6 metres is recommended for small farms. In case of tractor powered implements are to be used, width greater than 6 metre may be adopted. The excess water is allowed to flow through small furrows to a nearby tank. If tank is not available bunds may be made within fields at appropriate intervals. Red soils are more prone to erosion and have low water holding capacity. The runoff in these soils prior to primary tillage when the soil was compact and crusted was of the order of 50-70 per cent which reduced to 25-35 per cent after the field was cropped. On flat lands, therefore, appropriate primary tillage with field bunding may be suitable enough to check the erosion. For lands with 1-3 per cent slope, a method of graded border strip with a gradient of 0.2-0.4 per cent and length 100-150 m is most efficient way to conserve the moisture (Channappa *et al.,* 1988). For slopes more than 3 per cent and upto 6 per cent broad base terraces with a vertical interval<=0.5m and inward slope should be used for cultivation. Vegetative barriers may be used in place of terrace bunds. On very marginal lands such as red soil with gravelly structure and with slope 1-4 per cent, large field bunds of 60-75 cm size are made across the slope with both ends closed. The interspaces of these bunds are conveniently cultivated with good results. Seed yield of horsegram under different moisture regimes and mulches in semi-arid condition of West Bengal was studied during 1983-85. The crop was irrigated with 50 mm of water at the vegetative stage, flowering or at both the stages and given no mulch or about 5 tonnes leaf or straw mulch/ha. The seed yields were 8.4, 11.3 and 12.3 q/ha when irrigated at vegetative stage, flowering or both the stages, respectively. Application of straw mulch gave the highest seed yield of 12.5 q/ha followed by leaf mulch (10.5 q/ha) compared to control (8.6 q/ha). Further, the application of straw mulch together with two irrigations gave the highest level of productivity (14.2 q/ha) (Zaman and Mallick, 1991). Irrigation with 5 cm water per irrigation at cumulative pan evaporation ratio of 0.5 gave the highest yield of 19.5 q/ha on acid laterite soils. Yields were lower in other irrigation treatments (Lenka *et al.*, 1989).

Efficient Agronomic Management

Seedbed Preparation

Proper and timely tillage operations not only increase the moisture storage in the soil but also reduce the occurrence of weeds in the field. The effect of primary tillage is more pronounced on red loam, black cotton and acid laterite soils because of the tendency to get hardened at the surface after a long dry spell. On these soils one ploughing (20 cm deep) with mould broad or disc harrowing is recommended in kharif season. In the areas where horsegram crop is sown late in August after the rice or ragi, tillage with disc harrow may be done prior to sowing the crop. On light sandy soils, one tillage with sweep cultivator single type cultivator (15 cm deep) is sufficient. After each rain, hoeing up to 2-3 cm by a wheel hoe or hand hoe may help in moisture conservation.

Optimum Seed Rate

The effect of different spacings and seed rates on the growth and yield of horsegram was worked out. The seed yield of horsegram was unaffected by different spacings and seed rates at Konkan region of Maharashtra. However, mean yield was higher when sown at 30 cm row spacing, using 40 kg seed rate per hectare (Birari and Deorukhakar, 1983). In Bihar plateau region, a combination of 50 kg seed rate with 20 cm row spacing proved the best, giving a seed yield of 20.6 q/ha. Seed rate of 40-50 kg/ha gave significantly higher yield over 20 kg/ha. Further, there was no significant difference with seed rates of 30 to 40 kg per hectare (Rafey and Srivastava, 1989).

Optimum Planting Time

Horsegram is sown from first fortnight of June in Himachal Pradesh to kharif season in Rajasthan, M.P., parts of Maharashtra to late kharif in Konkan, Orissa, Bihar Plateau region, Karnataka, A.P. to winter season (Oct.-Nov.) in Tamil Nadu. In Himachal Pradesh, seed yield decreased from 10.6 q/ha to 4.6 q/ha with the delayed sowing dates from 15 June to 30th July (Thakur and Sharma, 1992). In Rajasthan, poor yields were realized with late sown crop (Henry and Daulay, 1988). In M.P., the planting of horsegram beyond 28th August reduced the seed yield (Bajpai *et al.*, 1990). In Bihar plateau region, horsegram is normally planted during third week of August to first week of September (Rafey *et al.*, 1988). It is indicated that the crop sown on 24th August gave significantly higher seed yield (12.3 q/ha) and there was drastic reduction in grain yield following advanced sowing time. In Maharasthra at Solapur, different horsegram varieties (*viz.*, D 40-1 and K-42) were grown during 1983-87 under rainfed conditions to find out optimum sowin time. It was indicated that both varieties could be sown from second fortnight of June to first fortnight of July. The var. K-42 performed better under late sowing condition up to first fortnight of August (Shinde *et al.*, 1993). In late sown conditions of Karnataka, horsegram (cv. BGM-1) gave average yield of 8.2 q/ha followed by cowpea (11.0 q/ha) being higher to soybean, fieldbean and red gram under late sown conditions in the *kharif* season (Yadahalli and Jayaram, 1975). In another study at Bangalore during 1974-75 to

1981-82, it was indicated that sowing of horsegram up to middle of August gave higher yield and delay in sowing up to first fortnight of October resulted in significant reduction in yield excepting during 1977-78, due probably to late rains (Yadahalli and Siddappa, 1983). It was found that seeds from August sown crop in Bangalore situations, were shriveled, discoloured and exhibited poor germination and affected by fungal infection because of rains in November during pod maturation (Rangaswamy *et al.*, 1991). In Tamil Nadu, it was observed that sowing done during second fortnight of October was found better (Nadanam, 1983).

Inter and intra-row spacing:

In general, the photo-sensitive varieties require wide space for early sowing and closer spacing for late sowing. At Dharwad, higher seed yield was obtained at 30 or 37.5 cm row spacing over 22.5 cm inter row spacing (Chandranath and Hosmani, 1995). At Bangalore, long duration varieties like BGM-1 gave higher yield (10.4 q/ha) with row spacing of 37.5 cm during middle July sowing. Comparable yields were obtained with short duration variety HH-2 (10-1 q/ha) with 30 cm row spacing. Late planting gave higher yields with closer spacing (Yadahalli and Siddappa, 1983). A closer row spacing (15, 22.5 or 30 cm) gave higher yield but resulted in reduction of some yield parameters (Nagaraju *et al.*, 1998). In Tamil Nadu, decreasing row width from 35 to 30 cm and 25 cm increased the seed yield of 3 horsegram varieties while the same decreased in case of Co-1. Yield was highest (708 kg/ha) in var. HPK-5 (Thirumurugam *et al.*, 1981). The higher yield at closer spacing of 25 cm between rows and 10 cm between plants was also reported by Nadanam (1983). In Maharashtra, dry sowing of horsegram done on 11, July increased the seed yield by about 120 per cent compared to sowing done on optimum soil moisture conditions after the onset of monsoon rains (Patil *et al.*, 1981).

Integrated Nutrient Management

Green manuring/residue management: The finger-millet grown after horsegram as green manure crop produced significantly higher seed yield (11.5 q/ha) over sole fingermillet at Bangalore (Singh and Venkateswarlu, 1985). Effect of various phosphatic fertilizers on fertility improvement in combination with lime and nitrogen given to the first crop of sunhemp grown on lateritic acid soils of West Bengal under reclamation was worked out on other crops. After ploughing sunhemp, the residual P^{+} lime markedly increased the seed yield of sesamum in the second year and of sunhemp for green manure in the third year (Sarkar, 1973). Horsegram is reported to utilize the residual soil moisture and nutrients left by harvest of wet season crop better when it followed cultivation of non-legume crops (Jena and Mishra, 1990).

Fertilizer use: Effect of phosphorus deficiency on root growth, phytomass production and nutrient content of horsegram indicated that P deficiency decreased growth and biomass of the whole plant. Adverse effects were more pronounced on leaf growth. P deficiency completely inhibited production of tertiary roots, increased the length of primary and secondary roots but decreased their weight. Increase in P rates improved the growth and total uptake of N, P, K, Ca and Na by plants. Among the P sources, nitrophosphate followed by super phosphate were the best in acid hill soil of Palampur (Marwaha, 1982). Application of 40 kg P_2O_5/ha in early sown crop gave the highest seed yield in H.P. (Sharma, 1987; Thakur and Sharma, 1992). In other study on the Hill soils during 1988 and 1989, grain and straw yields were increased with the application of nitrogenous fertilizers (up to 20 kg N/ha) and phosphatic fertilizers (upto 60 kg P_2O_5 per hectare). Fertilizer (N, P, K) response of horsegram under All India Coordinated Agronomic Experiments was worked out at Mayurbanj and Koraput districts of Orissa during 1970-71 and 1971-72 in 23 trials. No response to N was obtained in Mayurbanj district, while at Koraput, appliation of 15 kg and 30 kg N showed almost the same order of responses of about 105-110 kg/ha. Phosphorus response in Mayurbanj was low but significant of 37 kg/ha to 60 kg P_2O_5/ha, while in Koraput the response was from 30 to 60 kg N/ha. To conclude, application of 30 kg N/ha + 60 kg P_2O_5/ha may give higher responses than application of either of the nutrients.

In working out the economics of fertilization, it is indicated that application of fertilizer to horsegram in Mayurbanj district was not found economical as it did not cover the cost of fertilizer applied. The results showed that the actual profit comes about Rs. 100/ha and this could be attained with application of low combination of N_{15} + P_{20} as compared to N_{30} + P_{60} or N_{30} + P_{60} + K_{60}. It can thus be concluded that application of N_{15} + P_{30} to horsegram was likely to yield the best profit (Mahapatra *et al.*, 1973). Acharya (1983) indicated that application of 20 kg P_2O_5/ha gave optimum yield and met its requirement of phosphorus in Orissa. Application of lime in the main crop of maize during *kharif;* the yield of horsegram taken as a second crop increased to the extent of 22 per cent (5.9 to 7.27 q/ha). In another study in Orissa, on the impact of different doses of N, P under rainfed as well as under irrigation was worked out. It was indicated that the application of N and P under rainfed conditions gave the yields in the range of 11.0 –11.3 q/ha; under irrigation it ranged from 15.3-17.5 q/ha compared to control yield (10.3 and 12.0 q/ha). Application of N+P as 10+20 kg/ha to rainfed crop and at 20 +60 kg/ha to irrigated crop were the most suitable doses. Under late sown conditions in Orissa, the application of 20 kg N+ 40 kg P_2O_5/ha gave the optimum seed yield (Patra, 1991). In Rajasthan, yield of both local and KS-2 varieties increased due to phosphorus application upto 8.8 kg P per hectare and was highest at 50 cm row spacings (Singh and Singh, 1992). In another study at Sumerpur (Rajasthan), the seed yield of var. KS-2 increased due to 10 kg N and 17.6 P but there was no, response to zinc application (Choudhary *et al.,* 1994). Appliation of 20 kg N was superior over control and 10 kg N/ha while application of phosphorus (10, 20 and 30 kg/ha) gave significantly higher seed yield over control, under late sown conditions of Bihar plateau region (Rafey and Srivastava, 1989). In another study during kharif 1990-92 at Ranchi, higher average yield of horsegram was obtained with N and P application. NPK appliation (10, 20, 20, or 20, 40, 40 kg/ha) increased yield over unfertilized controls (Dwivedi *et al.*, 1996). In study at Konkan, increasing N doses from 0 to 37.5 kg/ha increased the dry matter accumulation throughout the growth period and increased seed yields from 683 to 952 kg/ha. Dry matter accumulation with 20-60 kg P/ha was also similar at all phosphorus rates but it was higher than without phosphorus application. The 60 kg P_2O_5/ha application gave the highest seed yield (Verenkar and Thorat, 1988). In another study, horsegram responded to 40 kg P_2O_5 and 20 kg sulphur per hectare (Nagaraju *et al.,* 1998) in Karnatka situations. At Bijapur (Karnataka), application of 25 kg N/ha and 40 kg P_2O_5/ha gave significant increase in yield over untreated control. This increase was a consequence of significant improvement in the number of pods/plant, pod weight, seed weight per plant and 1000-seed weight. In Dharwad region, horsegram responded to the application of 30 kg P_2O_5/ha (Chandranath and Hosmani, 1995). The response of horsegram in sub-humid regions of Nigeria, the application of different doses of phosphorus indicated that the dry matter yield increased upto 80 kg P_2O_5/ha with delay in harvesting time. The seed yield also increased upto the phosphorus application of 80 kg P_2O_5/ha (Omokanye, 1996).

Micro-nutrients: The response of horsegram to the application of micro-nutrients in lateritic acid soil of West

Bengal indicated that application of Mo to soil increased the dry weight of stem, leaves and nodules, number of pods/plant and seed yield in pot culture but had no effects on root growth. Application of 100 g Mo/ha to soil was superior to its appliation as a seed treatment or foliar spray (Pradhan and Sarkar, 1985). The chlorotic leaves found to have lower content of Zn, Fe, Cu and Mn. Soil test indicated deficiency of Fe and Zn. Foliar spray of Fe (60 ppm) in the form of $FeSO_4$ partially removed the symptoms and increased the yield by 20 per cent. Application of 30 kg S/ha did not have any significant effect on days to flowering, plant height, number of seeds per pod and 100-seed weight. However, significant difference were observed in respect of days to maturity, number of branches per plant, number of pods per plant and seed and straw yields only. The increase in number of pods per plant and seed and straw yields were 9.97, 7.54 and 7.00 per cent, respectively, over control. The B : C ratio was highest with 30 kg S/ha as evident with increase in seed yield. Similar findings have been reported by Dhankar *et al.* (1991) in pulses and Kumar and Pareek (1997) in cowpea. The profound influence of sulphur fertilization on growth and yield attributes might be due to increase in metabolic processes in plants which have promoted meristematic activities causing higher apical growth and photosynthetic surface.

Biofertilizers: Effect of rhizobium inoculation and phosphate appliation (22.4 kg P_2O_5/ha) was worked out at Sambalpur in Orissa for horsegram. Treatment with rhizobium inoculation alone could not have any effect in increasing the seed yield of horsegram. However, the number of nodules, N and dry matter contents of shoots and roots were significantly greater in treatment with rhizobium inoculation with P application and inoculation + P than without inoculation. The inoculated seeds along with phosphorus application resulted in 44.3 per cent increase of seed yield over uninoculated seeds (Sahu, 1973). At Bangalore, no response to rhizobium inoculation was observed (Yadahalli and Siddappa, 1983). However, in another study, inoculation in the absence of added nitrogen reduced grain yield by 13.7 per cent (Nagaraj Rao, 1982).

Effect of thiourea: Application of different treatments of thiourea at Pattambi and Bangalore stations on horsegram were studied from 1999 to 2003. The results reveal that 500 ppm thiourea seed soaking + 500 ppm spray one at vegetative and one at flowering stages gave highest seed yield (7.28 q/ha) over control (5.95 q/ha) and increase was 22.35%.

Reference

Acharya, N. 1983. Prospects and future of kulthi cultivation in Orissa. Paper presented in *Ist ICAR Group Meeting of Horsegram Research Workers* at AICRPDA, 14-15 Feb., 1983, Hyderabad.

Arya, M.P.S., Singh, R. and Singh, R. 1996. Effect of NPK on the nodulation, growth and yield characters of horsegram. *Legume Research,* 19: 65-69.

Bajpai, K.P., Singh, V.K. and Dwivedi, R.K. 1990. Performance of horsegram on different dates of sowing, phosphorus levels and intercropping. *Indian J. Pulses Res.* 3: 159-162.

Basavaraju, T.B. and Rao, M.R.G. 1997. Effect of *Acacia nilotica* and *Azachirachta indica* on the yield of horsegram in agri-silviculture systems, *Karnataka. J. of Agril. Sci.* 10: 1210-1212.

Bhatt, B.P., Chauhan, D.S. and Todaria, N.P. 1993. Phytotoxic effect of tree crops on germination and radicle extension of some food crops. *Tropical Sci..* 33: 69-73.

Birari, S.P. and Deorukhakar, N.V. 1983. Improvement of horsegram in Konkan. Paper presented in *Ist ICAR Group Meeting of Horsegram Research Workers* held on 14-15 Feb. 1983 at AICRPDA, Hyderabad.

Chandranath, H.T. and Hosmani, M.M. 1995. Effect of row spacings and phosphorus levels on phosphorus uptakes by horsegram, *Karnataka. J. Agril. Sci.* 83-86.

Chandrasekharan, N.R., S.V. Muhammad, N. Srinivasalu, N. Sundaram, P. Siva Subramanian and M. Rangaswamy 1974. Rotation experiments with sesamum. *Madras Agril. J.* 61: 505-509.

Channappa, T.C., R. Ramaiah and K. Ranga 1988. Watershed development and management for red soil of Karnatka. *In Proceedings of Third National Convention of Agricultural Engineers on "Watershed Development and Management"* held on Feb. 6-7, 1988 at Coimbatore.

Choudhary, L.S., Ishwar Singh and I. Singh 1994. Effect of nitrogen, phosphorus and zinc on growth and yield attributes of rainfed horsegram, *Indian J. Agri. Sci.* 64: 257-258.

Dhankar, J.S., Kumar, V. and Karwasra, S.P.S. 1991. Effect of S application on dry matter yield of horsegram (*Dolichus biflorus* L.) and its agronomical apprisal. *Indian J. Agron.* 28: 432-437.

Dwivedi, D.K., A. Kumar, S.S. Pandey and A. Kumar 1996. Response of horsegram cultivars in different fertility lands under rainfed conditions *J. Applied Biol.* 6: 70-72.

Henry, A. and Daulay, H.S. 1988. Genotype x environment interactions for seed yield in horsegram. *Indian J. Agric. Sci.* 58: 794-795.

Jadhav, A.S. and R.K. Rathod 1990. Studies on Marvel-8 (grass) + field crops intercropping on shallow soils. *J. Maharashtra Agril. Univ.* 15: 192-195.

Jena, D. and C. Mishra 1990. Symbiotic N fixation and fertilizer nitrogen use efficiency in legume-cereal intercropping systems. *J. Indian Soc. Soil Sci.* 38: 667-673.

Jha, L.K., G. Jha and K.K. Srivastava 1989. Selection and evaluation of suitable tree species and food crops under agro-forestry systems. *In Environmental Assessment and Management of Social Forestry in Tribal Regions Proc. of 3rd Mendelian Soc. of India,* held at Birsa Agril. Univ. 4-7 April 1989 at Ranchi.

Kambar, N.S. and S.N. Kadappa 1983. Review of plant improvement and agronomic research on horsegram in north Karnataka. Paper presented in *Ist ICAR Horsegram*

Research Workers at AICRPDA, 14-15 Feb. 1983, Hyderabad.

Kumar, S. and Pareek R.G. 1997. Effect of S on productivity of cowpea. *Ann. Agril. Res.* 18: 518-19.

Kumar, S. and Pareek, R.G. 1997. Effect of S on Productivity of cowpea. *Ann. Agril. Res.* 18: 518-519.

Lenka, D., A.K. Behura and R.K. Nayak 1989. Effect of irrigation and fertilization on yield of horsegram and uptake of nitrogen and phosphorus in acid laterite soils. *Orissa J. Agril. Res.* 2: 10-15.

Mahapatra, I.C., S.R. Bapat, M.G. Sardana and N.L. Bhendia 1973. Fertilizer response of wheat, gram and horsegram under rainfed conditions. *Fertilizer News* 18: 57-63.

Marwaha, B.C. 1982. Efficiency of phosphatic fertilizers differing in water solubility, their effect on growth and nutrients uptake by horsegram on acid alfisols. *Proc. Ind. Nat. Sci. Academy* 48: 434-439.

Mishra, R.K., S.K. Choudhary and A.K. Tripathi 1997. Intercropping of cowpea and horsegram with sorghum for fodder under rainfed conditions. *Ind. J. Agronomy,* 42: 405-408.

Nadanam, J. 1983. Agronomic research on horsegram in Madurai and Pudukkottai regions. Paper presented in *Ist ICAR Group Meeting on Horsegram Research Workers*, 14-15 Feb., 1983 at AICRPDA, Hyderabad.

Nagaraja, Rao, N. 1982. Studies on the effect of inoculation and nitrogen levels on the growth and yields of horsegram. *M. Sc. (Agric.) Thesis*, Univ. Agril. Sci., Bangalore.

Nagaraju, A.P., M.Y. Chalapathi, Abdul Khalak, K.P. Vishwanatha and Khalak, A. 1998. Response of horsegram to phosphorus and sulphur fertilization. *Crop Research* 15: 140-145.

Nayak, B.L. 1993. Performance of horsegram grown as a winter crop wider double cropping system in rainfed upland latorol Soil. *Orissa. J. of Agril. Res.* 6: 107-112.

Omokanye, A.T. 1996. Performance of horsegram in the sub-humid zones of Nigeria. *Legume Res.* 19: 52-54.

Patil, N.D., B.A. Koregave, S.S. Salunke, P.L. Narkhede, R.R. Walugkar and N.S. Bhadange 1981. Dry sowing of rainfed crops according to rainfall probabilities. *J. Maharashtra Agril. Univ.* 6: 76-77.

Patil, V.C., H.V. Nanjappa, B.K. Ramachandrappa and T.V. Muniyappa 1987. Effect of weed competition in drill sown fingermillet intercropping system. *Current Res.* Univ. Agril. Sci., Bangalore 16: 162-164.

Patra, S.S. 1991. Response of horsegram to dates of sowing and fertilizer under rainfed conditions. *Indian J. Agron.* 36: 296-297.

Pradhan, A.C. and A.K. Sarkar 1985. Effect of micro-nutrients on horsegram in lateritic acid soils of West Bengal. *Seeds and Farms* 11: 17-22.

Rafey, A. and V.C. Srivastava 1989. Agronomic management of late sown horsegram. *Indian J. Agron.* 34: 461-463.

Rafey, A., R.B. Prasad and V.C. Srivastava 1988. Studies on seeding time of horsegram. *Indian J. Agron.* 33: 335-336.

Rangaswamy, K.T., T.B. Anil Kumar and Yadahally, Y.H. 1991. Seed health of horsegram in relation to dates of sowing. *Pl. Protection Bull.* 43: 45.

Sahu, S.S. 1973. Effect of rhizobium inoculation and phosphate application on black gram and horsegram. *Madras Agril. J.* 60: 989-993.

Sarkar, A.K. 1973. Effect of various phosphatic fertilizers in combination with lime and nitrogen on the fertility improvement of lateritic soils. *Indian Agricult.* 17: 1-6.

Sharma, S.K. 1987. Grown attributes and macro-chemical constituents as indices of dates and graded phosphorus levels in horsegram. *J. Plant Nutrition* 10: 2031.

Shinde, S.K., C.B. Gaikwad and J.D. Patil 1993. Optimum sowing time for horsegram under dryland conditions. *Indian J. Agron.* 38: 436-438.

Singh, I. and G. Singh 1992. Response of horsegram varieties to phosphorus levels and sowing spacings. *Indian J. Agric. Sci.* 62: 400-01.

Singh, R.P. and J. Venkateswarlu 1985. Role of AICRPDA in research development. *Fertilizer News* 30: 46.

Singh, R.P., S.K. Das, U.N. Bhaskara and M.N. Reddy 1990. *Towards sustainable dryland agricultural practices.* CRIDA, Hyderabad, p. 20.

Singh, S.K. and K.R. Verma 1988. Allelopathic effects of leachates and extracts of *Pinus roxburghii* on four legumes in Kumaun Himalayas. *Indian J. Agric. Sci.* 58: 412-413.

Thakur, D.R. and S.K. Sharma 1992. performance of horsegram under different dates of sowings and phosphorus levels in rainfed mid hills. *Himachal J. Agril. Res.* 16: 12-14.

Tharmarajah, S.K. and H.M.W. Herath 1988. Studies on the pioneer cropping of dry Patana grasslands in the up country of Sri Lanka. *Tropical Agriculturist* 144: 99-100.

Thirumurugam, V., A. Chamy and A. Subramanian 1981. Response of horsegram varieties to spacings. *Pulse Crops Newsl.* 1: 44.

Verenkar, S.V. and S.T. Thorat 1988. A note on dry matter accumulation in horsegram as influenced by nitrogen and phosphorus fertilization. *J. Indian Soc. Coastal Agril. Res.* 6: 89-90.

Wali, B.M., S.B. Kalaghatagi and Y.B. Palled 1989. Intercropping in niger under rainfed conditions. *Indian J. Agron.* 34: 492-493.

Yadahalli, Y.H. and B. Siddappa 1983. Agronomic investigations on horsegram. Paper presented in *Ist ICAR Group Meeting of Horsegram Research Workers*, at AICRPDA, 14-15 Feb. 1983, Hyderabad.

Yadashalli, Y.H. and G. Jayaram 1975. Cowpea variety C-152 more profitable than horsegram under late kharif planting. *Current Research, Univ. Agril. Sci.* Bangalore 4: 114.

Zaman, A., and S. Mallick 1991. Water use and seed yield of horsegram under different moisture regimes and mulches in semi-arid region of eastern India. *J. Agronomy & Crop Sci.* 167: 39-42.

5

Plant Protection

S.M. Purushothaman, Mani Chellappan, E. Sreenivasan and K. Karthikeyan

Horsegram (*Macrotyloma uniflorum* Lam.) Verde is a hardy, drought resistant annual grain legume, grown extensively in Peninsular India as a poor man's pulse crop (Pursegloue, 1968). It can withstand adverse climate/ poor soils with low inputs. It is the cheapest source of dietary proteins, vitamins, calcium and iron. Horsegram is a rich source of urease and is recommended for patients suffering from urinary and kidney problems on account of its diuretic properties. The production potential of this drought hardy and poor men's crop can be furthered if proper plant protection umbrella is provided, particularly during infestation time. It is therefore, desired to study and review this aspect in details. This crop is infected by many diseases and pests. The important diseases and pests, which affect horse gram, are given below.

DISEASES

1. **Anthracnose** : *Colletotrichum lindemuthianum* (Sacc. & Magn.) Bri. & Cav.

Sub. Division: Deuteromycotina; Order: Melanconiales; Family: Melanconiaceae

It is a major disease of horsegram. It can also infect other pulse crops like cowpea, blackgram, beans etc.

Symptoms

The fungus may attack all plant parts above ground level at any stage of plant growth. The most prominent symptom is a characteristic spotting on the pods. Firstly, water soaked lesions appear on the pods, later becoming brown and enlarging to form circular spots of varying size. The spots are usually depressed with dark centers, and bright red, yellow or orange margins. They may occur more often on the under surface of the leaf than on the upper surface and also occur on the petioles and stems. When the infections are severe on the leaf petiole and stem the affected parts may wither off. Often the seedlings are blighted due to infection soon after the seeds germinate (Plate 1).

Pathogen

The mycelium is branched, septate, hyaline at first, becoming dark later with age. This will aggregate beneath the epidermis to form the stroma. Short conidiophores and setae arise from the fruiting structure called acervulus. The conidia are hyaline oblong, cylindrical with ends rounded or somewhat pointed at the end. They are single celled and have prominent oil globule. The fungus is carried on the seed and causes primary infection of the seedlings. Secondary spread may be due to air-borne conidia. The casual organism also lives on infested plant debris.

Management

Crop rotation and field sanitation should be adopted to eliminate the disease. Seed treatment with carbendazim @ 2g/kg of seed will reduce the primary inoculum.

The disease incidence in the field can be managed with carbendazim 0.1% spray (Thakur and Khare1990).

2. Powdery mildew : *Leveillula taurica* (Lev.) Arnaudi.

Sub. Division: Ascomycotina; Order: Erysiphales; Family: Erysiphaceae

The powdery mildew disease is also one of the important diseases of horse gram. It can be also seen in other pulse crops like cowpea, black gram, green gram etc.

Symptoms

A white powdery growth occurs on the leaves, spreading to cover the stem and other plant parts. The infection may take place at any stage of plant growth, but more severely when the plants are flowering and it persists until harvest. In very severe cases, the entire plant becomes affected and wilted.

Pathogen

The pathogen is enophytic and occupies spongy parenchyma of mesophyll. The haustoria penetrates into some of the parenchymatous cells. Conidiophores emerge in clusters through the stomata. The conidia are hyaline varying in size and shape; bear minute papilla like projection at the broader end, borne singly at the tip. They germinate readily in water producing a germ tube from one end. The fungus is air-borne, spreading from one field to another and from one host species to another in the vicinity, when favorable conditions prevail.

Management

Two sprays of calixin (0.05%) were found effective in controlling powdery mildew of horse gram with higher cost: benefit ratio (Ganesha Naik 2000). Since, this is not economical; attempts should be made to evolve resistant varieties to combat the disease.

3. Dry root rot: *Macrophomina phaseolina* (Tassi) Goud.

Sub. Division: Deuteromycotina; Order: Sphaeropsidales; Family: Sphaeropsidaceae

Dry root rot can infect horse gram at the time of flowering and pod formation stages. This disease can be seen in other crops like cowpea, black gram and green gram etc.

Symptoms

The first outward symptom of the disease is yellowing of the leaves. Within a day or two the leaves droop and in three or four days they drop off. The affected plants wilt and die within a week. The bark of the root and basal stem are fibrous and are found associated with black powdery mass of sclerotia of the fungus. The plants bear pods with partially filled seeds. The disease appears in patches and becomes severe during dry periods.

Pathogen

The fungus invades the host both inter and intra cellularly. It grows rather fast, covering large areas of the host tissues and eventually killing them in a short time. It produces numerous sclerotial bodies on the host tissue. Often the conidial or pycnidial stage is produced on the host. The fungus is a facultative parasite capable of living sprohytically on dead organic tissues, particularly on many of its natural hosts, producing sclerotial bodies. It produces the pycnidia when the atmospheric temperature is above 30°C and the pycniospores remain viable for over a year. The fungus is mainly a soil dweller and spreads from plant to plant through irrigation water, tools and implements and cultural operations. The sclerotia and pycniospores may also become air-borne and cause further spread of the pathogen. Low soil water potential favours production of large quantities of *M. phaseolina* sclerotia (Olaya and abawi, 1996).

Management

Application of organic amendments like farmyard manure, neem cake and seed treatment with *Trichoderma viride* @ 4g/kg of seed and *Pseudomonas fluorescens* @ 10 g/kg of seed can reduce the disease incidence. Enhanced soil moisture can reduce the *M phaseolina* population (Lodha, 1996).

4. Leaf spot: *Cercospora dolichi* Ell. & EVr.

Sub. Division: Deuteromycotina; Order: Moniliales; Family: Dematiaceae

The leaf spot caused by *Cercospora* can be noticed in other pulse crops also.

Symptoms

Usually water soaked lesions appear on the leaf blade and soon the affected tissues turn brown to reddish brown well defined spots often bound by veins and purplish border develop. The centre of which may turn gray. Powdery growth, due to the sporulation of the fungus, may also be seen at the centre of the spots. Though the spotting is mostly confined to leaf blades, but occasionally may occur on the fruit and floral parts. In some cases the affected tissue is killed and the dead portion drops out leaving shot-hole symptoms on the leaves (Plate 1).

Pathogen

The fungus produces short condiophores and multicelled conidia, which are narrow and long. The disease can spread through infected seeds and wind-borne conidia of the fungus.

Management

Seed treatment with captan at the rate of 3 g/ kg of seed can be used. Spraying the infected plant with mancozeb (0.2%) can reduce the incidence of the disease.

5. Rust : *Uromyces phaseoli typica* Arth. (Syn. *U. appendiculatus* (Pers) Fries.)

Sub. Division: Basidiomycotina; Order: Uredinales; Family: Pucciniaceae

The rust disease can be seen in other pulse crops also.

Symptoms

The fungus produces characteristic rust pustules on the plant. The pustules are mostly found on the leaf blade. They are more conspicuous on the under surface of the leaves. Often a number of such pustules may occur on the same leaf. The infection may spread to young stem also. Plants with heavy rust infection will give a brown tinge when looked from a distance. In advanced stage of infections the leaf may wither resulting in considerable damage to the crop.

Pathogen

The fungus is an obligate pathogen. The uredial stage of the fungus repeats several times. The urediospores build up in masses and are spread by wind to cause secondary infections. The fungus also perpetuate through the teliospores. The teliospores germinate to produce basidiospores, which produce in turn pycnia, aceia and uredia.

Management

Development of rust resistant varieties and adjusting the sowing dates of the crop to avoid severe infections of

Leaf spot disease

Podfly damage

Leaf hairy caterpillar damage

Spotted caterpillar damage

Yellow Mosaic Virus

Anthracnose disease

Plate - 1

Damage due to major diseases and insect-pests in horsegram

the plant in the field may help considerably in reaping better harvest.

6. Aerial blight: *Rhizactonia solani* Kuhn

Sub. Division: Deuteromycotina; Mycelia Sterilia

The disease can cause 10-60% loss depending upon the severity of the disease.

Symptoms

Aerial blight is seen on the foliage as irregular water soaked area. Under high atmospheric humidity the spots coalesce rapidly and cover a large part of the leaf lamina. There will be white mycelial growth also on the affected area. Severely affected leaves shed in large numbers.

Pathogen

The perfect state of the pathogen is called as *Thanatephorus cucumeris* (Fr.) Donk. The hyphae of the organism are colourless when young, turning yellowish brown on maturity. Sclerotia are superficial, more or less globose, but flattened below. They are white when young but turn dark brown. The disease spreads through infected soil.

Management

The management of the diseases caused by *R. solani* is difficult due to its saprophytic surviving ability in the soil. Use of soil fungicides will not be economical. Soil application of *Trichoderma viride*, *T. harzianum* and *Gliocladium virens* significantly reduces the mycelial and sclerotial production of *R. solani* (Dubey 1998).Use of resistant varieties will be the only economical control measure.

6. Yellow mosaic : Mung bean yellow mosaic gemini virus

Symptoms

Initially small yellow patches or spots appear on the green lamina. Young leaves first show the symptoms. In due course the area of yellow discoloration increases in new growths and may completely turn yellow. The infected plants mature later and bear very few flowers and pods. The pods are small and distorted. The seeds are also reduced in size and are shriveled (Plate 1).

Pathogen

The disease spreads through the insect vector *Bemisia tabaci*. The vector acquires the virus with a minimum acquisition-feeding period of 30 minutes. But, it requires another 6 hours to become viruliferous and a minimum inoculation-feeding period of 10 minutes. The whiteflies remained infective for up to 12 days. A single whitefly can transmit the yellow mosaic virus (Muniyappa and Reddy 1976). The symptoms appear in 7-12 days. *Indigofera hirsuta* acts as natural reservoir of horse gram mosaic virus (Muniyappa and Reddy 1976). This virus can also infect *Vigna mungo*, *Vigna radiata*, *Cajanus cajan* and *Vigna aconitifolia* etc. (Bikash Mandal and Anupam Varma, 1996).

Management

Use of resistant cultivars for cultivation is one of the best strategy of disease management, removal of the infected plant and periodical spray with systemic insecticide monocrotophos (0.1%) or neem oil soap emulsion 3% spray (Sethuraman *et al*., 2001) and early sowing during first fortnights of August and September (Naik, 2001) can minimize the disease incidence.

Some of the other diseases infecting horse gram crop are wilt (*Neocosmospora vasinfecta*), leaf spot (*Pyreno-*

chaeta dolichi), and root rot (*Pellicularia filamentosa*) and die back (*Colletotrichum capsici*).

INSECT PESTS OF HORSE GRAM

1. Pod caterpillar: *Helicoverpa armigera* (Hb) (Family: Noctuidae; Order: Lepidoptera)

This polyphagous species is an important pest of pulses also. The adult moth has a V shaped specks on the light brownish forewings and a dark border on the distal end of hind wings. The adult lays sculptured spherical yellowish eggs with flattened bottom singly on tender leaves and shoot.

Symptoms of damage

The young larva feeds on the tender leaves, buds and flowers for a short time and subsequently it bores into the pods and feeds on the seeds with its head and part of the body thrust inside, the rest of dorsally on the body. The larva measures 35-45 mm long and greenish with dark grey stripes laterally on the body.

Management

Attract and kill the nocturnal adults using light trap or pheromone lure. Spray the crop with *Helicoverpa* NPV @ $1.5X10^{12}$ POB/ha.

2. American serpentine leaf miner: *Liriomyza trifolii* Burgess (Family: Agromyzidae; Order: Diptera)

Adult is a very small (1-1.5 mm) pale yellow fly. The eggs are trusted on the leaf surface.

Nature of damage

The emerging maggots mine the leaves and feed on the mesophyll making serpentine on the leaves. Due to severe mining by the larvae and feeding punctures by the adult flies leads to defoliation of crop. Several maggots mine on a single leaf and infestation is usually seen on mature leaves.

Management

The pest infestation can be reduced by collection and destruction of first affected leaves and by spraying 5% neem oil emulsion on the leaves.

3. Green leaf hopper: *Empoasca kerri* Pruthi (Family: Cicadellidae; Order: Hemiptera)

The adults are yellowish-green in color. Eggs are laid under side of the leaf, embedding them into the leaf veins. The nymphs are wedge shaped and very active more briskly forward and single ways. The winged adults are displaced at the slightest disturbances and are phototropic in nature.

Nature of damage

Nymphs and adults, generally confined to the underside of leaves, suck sap from veins and vein lets. As a result, tips of affected leaves become pale initially and later brown, turn upwards and get dried. During severe infestation brown spots appear on leaf margin, they get rolled and dried.

Management

Spray dimethate 30EC (0.05%) to check the population build up of the pest. The pest can also be managed by spraying the crop with imidacloprid @ 20 g ai/ha (Karthikeyan, 2003).

4. Pod bug: *Clavigralla gibbosa* S. & *C. horrens* D. (Family: Coreidae; Order: Hemiptera).

Nature and Damage

These active greenish brown bugs and nymphs suck juice of unripe seeds from the green pods. The attacked pods show light yellow patches on their surface and they shrivel and die subsequently.

Management

Spray with malathion (0.05%) is effective to reduce the pest population.

5. **Pod fly**: *Melanagromyza obtusa* M. (Family: Agromyzidae; Order: Diptera)

The adult is a small black fly with slight bright greenish tinge and glossy black abdomen. It lays eggs singly within pods by piercing the pericarp with its ovipositor. The eggs are needle shaped and can be seen projecting inwards from the walls of the pods. One pod may contain up to 5 eggs. The eggs hatches in about three days (Plate 1).

Nature of damage

The maggots first make galleries just below the epidermis of the seeds feeding deeper into the seeds later on. The larva pupates in the larval groove. The maggot causes damage to the green pods mining the seeds. The damaged grains are unfit for seed or consumption. The affected pods appear pale and undersized. One seed is enough for the development of the maggot. Adult fly emerges by breaking the membrane on the exit hole made by the larva before pupation and thus leaving the characteristic pinhead size holes on the pods.

Management

Attract and kill the fly using fish meal trap or spray carbaryl 50 WP @ 0.2% or dimethoate 30EC @ 0.05% in the early stages of the attack.

6. **Thrips:** i. *Megalurothrips usitatus* (Family: Thripidae; Order: Thysanoptera)

Thrips infest the flowers leads to abnormality and flower shedding before opening. Even if any such flower sets pod, that will be abnormal.

ii. *Frankliniella* spp. (Family: Thripidae; Order: Thysanoptera)

The nymphs and adults of the thrips feed on the leaves in all the stages of crop growth and cause crinkling of leaves in cases of severe attack.

Management

Spray applications of neem oil 5%, phosalone 0.05%, malathion 0.1%, dimethoate 0.03 per cent, methyl parathion 0.025 per cent etc. at the bud formation stage for the *M. usitatus* and as per the severity of the *Frankliniella* spp. were found effective against the pests.

7. **Whitefly**: *Bemisia tabaci* (Genn.) (Family: Aleyrodidae; Order: Hemiptera)

It is highly polyphagous and breeds on a variety of plants such as hollyhock, lady's finger, tobacco, safflower, *Achyranthus aspera*, Lab-lab, niger, tapioca etc. The females lay stalked eggs singly on the underside of the leaves. In horse gram the nymphs are found in large numbers on the under surface of the leaves and drain sap. It acts as a vector for the yellow mosaic disease in horse gram.

Management

Spray the crop with imadacloprid @ 40 g ai/ha.

8. **Spotted pod borer** : *Maruca testulalis* Geyer. (Family: Pyralidae; Order: Lepidoptera)

The moth has a white cross band on the dark brown fore wings and a border on the white hind wings. It lays eggs on or near the flower buds.

Nature of damage

The greenish caterpillar enters the bud, flower or the pod. With in the pod, the caterpillar feeds on the seeds resulting in appreciable loss in yield. The entrance hole is plugged with excreta. The larva may bore into the tender stems as well. The infested flowers and pods maybe webbed together. It pupates within debris or near surface of the ground.

Management

Spray the crop with carbaryl 50 WP @ 0.2%.

8. **Leaf roller**: *Nacoleia vulgaris* Guen. (Family: Pyralidae; Order: Lepidoptera)

The moth is slender with yellowish brown. The moth lays flat scale-like eggs in rows on tender leaves.

Nature of damage

The larva feeds on green matter of the leaves in its early stages. Later it webs together the leaves and feeds from within skeletonizing them completely. The damage caused in this way to the host plants is very serious. The pest attacks other pulse crops like cowpea, black and green gram during August to February. The caterpillar is parasitized by *Cardiochila fulvus* Cam. and *Xanthopimpla punctuator* L. and by a tachnid (Mammen and Jopesh, 1975).

Management

Remove and destroy the rolled leaves which contain different stages of the pest.

9. **Hairy caterpillar**: *Spilosoma obliqua* W.& *Pericallia ricini* Fb. (Family: Arctiidae; Order: Lepidoptera)

Nature of damage

The brown and black caterpillars feed on the leaves, soft stems, branches and may totally denude the crop within few days (Plate 1).

Management

Gregarious young larvae can be destroyed by pulling out the infested plants and burying them in the ground can minimize the population of the pest.

Some of the other pests infesting horse gram are *Monolepta signata* (Chysomelidae, Coleoptera) attacks the crop during seedling stage. *Etiella zinckenella* (Phycitidae, Lepidoptera), *Omiodes indicata* (Pyralidae, Lepidoptera), *Bagrada cruciferarum* (Pentatomidae, Hemiptera) and *Nezara viridula* (Pentatomidae, Hemiptera) affect the crop at pod maturity stage. *Aticarsia irrorata* (Noctuidae, Lepidoptera) affects during vegetative stage (Singh *et al.*, 1997), gross hopper *Colemania sphenareoides* (Saha, 1995) and mite (*Tetranychus neocaledonicus*) Andre. (Fotedar, 1978).

References

Bikash Mandal and Anupam Varma. 1996. Differentiation of natural variants of mungbean yellow mosaic geminivirus by host reactions and DNA-DNA hybridization. Intern. *Jour. Tropi. Pl. Dis.* 14 (2):189-202.

Dubey, S.C. 1998. Evaluation of fungal antagonists of *Thanatephorus cucumeris* causing web blight of horse gram. *Jour. Mycol. and Pl. Path.* 28(1): 15-18.

Fotedar, R. 1978. Records of a few phytophagous mites from Hisar, Haryana. *Acarology Newsl.* 7 (4):2 ref.

Ganesha Naik and Rangaswamy, S.D. 1994. Effect of date of sowing and application of chemicals on the incidence and control of powdery mildew in horse gram (*Dolichos biflorus*) Roxb. *Legume Res.* 17(4): 203-206.

Ganesha Naik. 2000. Chemical control of powdery mildew in horse gram (*Macrotyloma uniflorum* Lam.) Verdec. *Legume Res.* 23(4): 262-264.

Karthikeyan, R. 2003. Bioefficacy of newer insecticides against leafhopper, *Empoasca motti* P. in bitter gourd. *M.Sc. (Ag.) Thesis*, Kerala Agricultural University, Vellanikkara, Thrissur. p. 102.

Lodha, S. 1996. Influence of moisture conservation techniques on *Macrophomina phaseolina* population, dry root rot and yield of clusterbean. *Indian Phytopath.* 49 (4):342-349.

Mammen, K.V. and Joseph, K.V. 1975. Biology of the pea leaf roller, *Nacoleia vulgaris* Guence. *Agri. Res. J. Kerala.* 3: 51-54.

Mishra, D. 1988. A new wilt disease of horse gram in India. *Curr. Sci.* 57 (16): 898.

Muniyappa, V. and Reddy, H.R. 1976. Studies on the yellow mosaic disease of horse gram (*Dolichos biflorus* Linn.) I. virus vector relationships. *Mysore Jour. Agricu. Sci.* 10(4): 605-610.

Muniyappa, V. and Reddy, H.R. 1979. *Indigofera hirsuta* a natural reservoir of horse gram yellow mosaic virus. *Madras Agri. Jour.* 66(5): 350.

Naik. R.G. 2001. Effect of date of sowing on disease incidence and yields of horse gram (*Macrotyloma uniflorum*) Lam verdec. *Legume Res.* 24(3): 182-185.

Nair, M.C. and Menon, M.R. 1983. *Diseases of crop plants of Kerala*. Kerala Agricultural University Press, Mannuthy. p. 277.

Nair, M.R.G.K. and Visalakshi, A. 1978. *A monograph on crop pests of Kerala and their control*. Kerala Agricultural University Press, Mannuthy. p. 227. (Third revised edition 1999).

Olaya, G. and Abawi, G.S. 1996. Effect of water potential on mycelial growth and on production and germination of sclerotia of *Macrophomina phaseolina*. *Pl. Disease* 80: 1347-1350.

Purseglove, J.W. 1968. *Tropical crops, dictyledons*. Longmans Green and Co. Ltd., London and Harlow. p. 264.

Rangaswami, G. 1993. *Diseases of crop plants in India*. Prentice-Hall of India, Pvt. Ltd, New Delhi.

Saha, L.R. 1995. Hand book of plant protection. Kalyani Publishers, New Delhi. p. 213-214.

Sethuraman, K., Manivannan, N. and Natarajan, S. 2001. Management of yellow mosaic disease of urdbean using neem products. *Legume Res.* 24 (3): 197-199.

Singh, T.V.K., Venkateswarlu, O. and Sudhakar, R. 1997. Insect pests of horse gram, *Macrotyloma uniflorum* (Linn.) in Andhra Pradesh. *Insect Envior.* 3(2): 53-54

Srivastava, L.S. and Gupta, D.K. 1989. Aerial blight of french bean, ground nut, soybean, black, green and horse gram – a new record from India. *Pl. Disease Res.* 4(2):163-164.

Thakur, M.P. and Khare, M.N. 1990. Efficacy of different fungicides on the severity of mungbean anthracnose. *Indian Jour. Pulses Res.* 3 (1):94-96.

Vasantharaj David, B. and Kumaraswami, T. 1975. Elements of economic entomology. Popular off-Sets Saidapet, Chennai. p. 536. (Reprint 2000).

6

Quality, uses and Nutrition of Horsegram

K.P. Viswanatha, A. Henry, S.B.S. Tikka and D. Kumar

Horsegram a multipurpose crop is a cheap source of food, feed, fodder and manure. The popularity, of this legume is reckoned for its other qualities like, early culture, wide adaptation, drought resistance and use in soil conservation. The nutritional components of horsegram are no less important than any other legume crop. It is essential source of protein and minerals. Horsegram have high amount of calcium, phosphorus, iron and vitamins. It has good foliage which adds to the soil enrichment and increases its organic matter. It also has better nitrogen fixing ability. It is helpful in treatment of kidney and gall bladder stone. The grain is used as a human food and also a concentrated feed for cattle. Horsegram seed is rich in vegetable protein. It is ideal supplimentry diet in cool months of the year. Besides, it suffers from certain anti-nutritional factors, hence is known as poor man's pulse available at cheaper price than rest of the pulses. Horsegram seeds may be separated into grey, dark brown or black testa colours and into seeds, which either float or sink in water. Most of the black seeds float, whereas only few grey and dark seeds float. Per cent germination was about 94 per cent in grey and dark grown seeds and 56 per cent in black seeds. These differences were associated with 84 per cent germination in seeds which sunk and only 16 per cent germination in floating seeds (Omokanye, 1995). Information compiled on certain uses, nutritional, anti-nutritional and quality traits of horsegram seeds have been documented below:

Fatty acids

Horsegram seeds contain 1.1 per cent oil content (Ahmed *et al.,* 1986). The nature of horsegram lipids (1.7-2.2 per cent) was studied for component fatty acids and it was indicated that neutral lipids constituted 46-52 per cent, phospholipids 35-40 per cent and glucolipids 10-12 per cent (Mahadevappa and Raina, 1978).

Antinutritional factors

Lectin was detected in the cotyledons and the embryos of horsegram seeds but it was absent in leachate and testa. However, the leachate caused haemolysis, indicating the presence of some toxins (Kaushik, 1984). Proximate nutrients, calcium and some anti-nutrients were estimated in 16 varieties of horsegram on whole seed and dehulled seed basis. The protein, fat and carbohydrate contents were higher in the dehulled samples than in whole seeds. However, the moisture, fibre, ash and Ca contents of the dehulled samples were lower. A significant portion of the anti-nutrients were removed by dehulling (Sudha *et al.*, 1995). The formation of calcium oxalate, crystals and changes in the fine structure of crystal cells in the testa of the seeds during ripening were described by Solbrig (1983). The chemical composition of 3 vars. (black, mixed, yellow) of horsegram after roasting or soaking and sprouting indicated that sprouting had a beneficial effect on the crude protein, other extract and crude fiber content while an adverse effect on the N_2 free extract. The presence of antinutritional factors may limit the usefulness of these seeds for animal feed (Arora, 1997). The slow digestion in horsegram is attributed to the presence of Tripsin inhibitor. The effect of temperature on ascorbic acid content of sprouting horsegram seeds, i.e. low temperature sprouting (4-5^{0}C) at room temperature sprouting (27-28^{0}C) and high temperature sprouting (33-34^{0}C) indicated that sprouting was faster at room temperature and high temperature. Ascorbic acid values initially increased and then decreased. The number of days for which content increased was decreased with increasing temperature. The decline in ascorbic acid levels began earlier in the cotyledons than the embryonic axis (Madhurima Dikshit and Dikshit, 1992).

Protein and Mineral Nutrition

A number of reports on nutritive value of horsegram are available in the literature. Begum *et al.* (1977) have reported that the mean protein content of 96 collections was averaged to 23.12%. Patil and Deshmukh (1985) found the range of protein content in 50 cultivars from 17.9 to 28.8 per cent, the highest being in cv. White Shimoga and the lowest in EC-7460. Pore (1979) estimated the protein in the range of 14.8 to 26.9 per cent. Recent AICRP reports indicated that the protein content of horsegram seeds ranged from 27.45 to 31.61% and carbohydrate from 48.2 to 55.0% (Anonymous 2002). Evaluation of 39 genotypes of horsegram done at Sardar Krushinagar for protein content revealed range from 17.63 to 23.64%. Highest protein content was recorded in AK-21 (23.64%) followed by GHG-40 (23.61%) and GHG-38 (23.40%) (Anonymous, 2004). Other reports also indicate nutritive value of horsegram as: 18.40% digestible crude protein, 69.75% starch equivalent and 70% total digestible nutrients on dry matter basis. Pore (1979b) observed that among 27 improved varieties of horsegram, total mineral content varied from 2.2 to 3.3%, other constituents were: fat content 0.05-0.95 per cent, carbohydrate content 69-81 per cent, Ca content 191-430 mg/g, P content 244-495 mg/g and Fe content 13-70.6 mg/g. The other chemical constituents were, ash 2.4-3.6 g/100 g, flour (mean 3.2 g), Ca 191-430 mg (mean 285), Na 40.2-340 mg (mean (mean 101.5), oxalic acid 182-449 mg (Pore 1979b). The Ca: P ratio varied from 1:1 to 1.1:8. All the varieties had high amount of calcium and phosphorous and also contained good proportion of iron. Satpure *et al.* (1995) have reported that gamma rays induced mutations in horsegram had improved contents of N, Ca, K and Fe in grain. Satpute et al. (1996) indicated higher protein content in mutants ACCK and ACCL-302 of var. DPL-1 (21.9-21.5%). Murthy *et al.* (1980) estimated the protein content in the 20 varieties of horsegram from 22.02 to 31.16%. Peshin and Singhla (1994) observed a marked decrease in anti-calcifying activity with the maturation of seeds or post harvest storage for six months and found that the inhibitor of crystallization present in the seed extract were water soluble, heat stable, polar, non-tannin and non-protein in nature. Sudha *et al.* (1995) studied 16 varieties of horsegram and found that their dehulled seeds had higher protein, fat and carbohydrates content than in the corresponding whole horsegram. The moisture, fiber, ash and calcium contents of the dehulled samples were lower. Significant portion of the anti-nutrients studied were removed by dehulling. Mandal *et al.* (2004) reported quality traits of one wild horsegram accession (IC212722) collected from remote village Sar of Garhwal Himalaya in Uttaranchal, it registered very high seed protein value to the extent of 38.35%.

Horsegram is one of the cheapest sources of protein for both human beings and animals. It is also a cheap source of vitamins, calcium and iron (Katiyar, 1984); and also of urease. The average chemical composition of horsegram has been studied and reported below:

Constituent	Value
Water	8.4 %
Protein	24.7 %
Carbohydrate	59.8 %
Ash	2.8 %
Fats	4.8 %
Calcium	0.34 %
Phosphorus	0.27 %
Iron	0.02 %
Vitamin A	40-199 i.u./100g
Nicotinic acid	1.5 mg / 100 g
Thiamine	0.40 mg / 100 g
Riboflavin	0.15 mg / 100 g

Source: Ray, 1970

Mineral	Value (Mg / 100 g)
P	310
Mg	135
Ca	105
Fe	11.9
Mn	15
Zn	3.4
Cu	5.5

Horsegram is reported to contain higher amount of iron and molybdenum than any other legume generally consumed in India.

The amino acid content of horsegram has been studied as below follows:

Aminoacid	(Mg / g N)	(Mg / g N)
Isoleucine	420	344
Leucine	490	506
Lysine	490	537
Phenyl alanine	430	419
Tyrosine	160	331
Sulphur-aminoacids	200	- -
Methionine	90	106
Cystime	140	62
Threonine	240	212
Valine	370	356
Arginine	- -	519
Histidine	- -	156

Source: Ray, 1970.

The sulphur containing aminoacids cystine and methionine account for the lowest scores. The percentage of tryptophan and total sulphur amino acid are also lower in horsegram than in other pulses. However, total mineral content is comparable with other food legumes. Horsegram has fairly good amount of phosphorous. In addition, there is evidence that horsegram seeds contain growth-inhibiting factors, which can be inactivated by autoclaving (Ray, 1969).

Fodder quality

The biomass of the crop, which is highly nutritious, is used as fodder for cattle and buffaloes. Horsegram hay commonly known as bhusa, which is generally fed to cattle, is also very nutritious and highly palatable.

Chemical analysis of the hay on dry weight basis gave the following per cent values.

Content	Percentage
Crude protein	10.6
Crude fibre	16.2
Fat	1.80
N free extract	58.3
Ash	13.1
Calcium	1.81
Phosphorus	1.18

Uses

Horsegram grain is utilized for human consumption while the haulms serve as valuable cattle feed. Although not valued so much as the other pulses it is equally nutritive. It is largely eaten by the poorer classes, either after boiled or in the form of a meal variously prepared. This gram is a poor man's pulse since it is easily available to the rural folk at lesser cost than the other grain legumes. It is considered by the poor people as an ideal supplementary diet in the colder months of the year. Horsegram is consumed both as boiled and fried preparations. Alternatively it is also sprouted and then boiled for consumption. Unlike most other Asian grain legumes, it is not converted into split dhal/pulse. Roasted gram is ground with salt, chilly, sometimes with coconut and consumed with rice and millets ('Chapati). In south India a variety of products are made out of horsegram. It is made into curry (*'bassaru'in kanada*), *sundal* (*usli* in kanada) and also sprouted grain are mixed with greens and used as side dish. Some examples are given below:

Curry and usli : Grain is boiled like other *dhals* and the split pulse is used for making *sundal.* The extract which is known as *gravy* is mixed with *masala* and thus *curry* is prepared. This *curry* is consumed along with ragi *kali* by the farmers.

Pappad: It is also made commonly in Karnataka. The grains are soaked overnight [12-18 hours], dried thereafter, made into flour. Horsegram flour is mixed with blackgram flour followed by addition of *masala* ingredients like chilli powder, omam, salt, lime juice, etc., for making 'pappad'. This 'pappad' is not fried but roasted in hot cake, like dry *chapati*.

Sprouted horsegram is also mixed with greens and used as side dish. Certain communities are fond of horsegram gravy. The consistency of gravy is made viscous by boiling the grains. This gravy is retained even upto three days. Most of the moisture is made to evaporate by intense boiling, quality of this preparation is therefore, maintained.

In Mayanmar the seed is boiled, pounded with salt and fermented to produce a sauce similar to soyasauce. Seeds are stringent and well known for their medicinal (diuretic) properties. It is said to be good for patients suffering from urinary and kidney problems (Thakur, 1979).

Grains are also largely used as feed for cattle and horses, the working animals consume almost 1.5 – 2.0 kg per day per animal. During olden days, horsegram grains were fed to racing and military horses and hence, the crop named as horsegram. The seed coat is hard to break and the animals pass it with their excreta quite unaltered. The grains after light boiling are fed to horses and milch animals. In some parts, soaked and ground horsegram is fed to milching cattle.

Medicinal Value

It is supposed to have medicinal value especially in dissolving stones found in the kidneys By consuming the soup prepared from sprouted horsegram, one can get better sleep. In ladies irregular monthly periods and urinary problems can be over come by consuming the grains daily. Horsegram pappad will reduce acidity which is becoming common now-a-day. Grains may also be useful in curing whooping cough. Constipation can be avoided by drinking a cup of horsegram soup(grain boiled water) with a small quantity of lemon juice and salt.

Horsegram seeds is used to treat kidney stones in India and is a constituent of the Ayurvedic medicine, cystone. The *in vitro* effects of seed extracts on crystallization of calcium phosphate, a major constituent of kidney stones were studied. The results revealed that there was marked decrease in the anti-calcifying activity with seed maturation and post harvest storage of seeds for six months. It was suggested that the inhibitors of crystallization present in the seed extract were water soluble, heat stable, polar, non-tannins and non-protein in nature. There appeared to two or more different inhibitors of calcium phosphate precipitation. The anti-calcifying activity of the aqueous extract was lost completely when it was treated with activated charcoal and the activity was not recovered or eluded by any solvent tested. In a study to understand anti-calcifying properties of horsegram. The boiled water extract of whole grain is given to the patient and used in many parts of the country for dissolution and prevention of urinary stones (Singla and Kanwar, 1985). It is one of the major ingredients of Ayurvedic medicine 'CYSTONE' prescribed commonly to the patients suffering from kidney ailments (Mukherjee 1984, Singhla and Kumar, 1973). To date, not much is known about scientific basis of the action of this decoction. The aqueous extract of this drug is known to inhibit the initial precipitation of calcium and phosphorus ions (Jethi *et al.,* 1983), which has been confirmed by clinical studies (Joshi *et al.,* 1988). It is also useful in treatment of piles, tumors, heart trouble, bronchitis, cough and cold and urinary diseases. Archana *et al.* (1994) reported that the inhibitors of crystallization present in the seed extract were water soluble, heat stable , polar , non tannin and non-protein in nature. There appeared to be two or more different inhibitors of calcium phosphate precipitation since the dialyate and dialysed fractions continued inhibitory activity, although it was more marked for the dialysed fractions. Horsegram is more effective in inhibiting the formation of calcium and phosphate precipitates than the B. *ligulata*. The combination of the two extracts was not active as the individual extracts (Garimella *et al.*, 2001).

References

Anonymous. 2004. A SDAU-GVT collaborative research project on horsegram and blackgram. *Annual progress report.*

Anonymous 2002. Annual Progress Report of AICRP on Arid Legumes. Table 44 & 45, pages 139-140.

Archana, P., Singla, S.K., and Peshin, A. 1994. Anticalafying properties of horsegram. *Indian Jour. Expt. Biol.* 32: 889-891.

Arora, N. 1997. A preliminary study of horsegram. *Indian Vet. J.* 74: 802-803.

Ahmad, R., I. Ahmad, A. Mannan, F. Ahmad and S.M. Osman 1986. Studies on minor seed oils. XI *Fettz. Seifen Anstrichmittel* 88: 147-148.

Begum M., Srihara, J. and. Hiremath, S.R. 1977. Varietal difference in protein of horsegram *(Dolichos biflours Linn.). Mysore. J. Agric Sci.* **11**: 521-524.

Garimella, T.S., Jolly, C.I. and Narayanan, S. 2001. *In vitro* studies on antilithiatic activity of seeds of *Dolichos biflorus* Linn. and rhizomes of *Bergenia ligulata* Wall. *Phototherap res.y*, **15(4)**: 351-355.

Jethi, P.K., Duggal, B, Sohta, R.S. Gupta, M. and Safat, I.B. 1983. *Indian J. Med. Res.* 78, 422.

Joshi, P.K. Patil, D.B. Sukumaran K, Parasania R. and Patel, M.R. 1988. *Indian J. Vet. Surg*, 9,18.

Katiyar, R. P. 1984. 'Kulthi' a promising pulse crop for Himachal Hills. *Indian Fmg.*, 34 **(9):** 31-35.

Kaushik, P. 1984. Lectin from different parts of the seeds of Dolichos proc. *Indian Nat. Sci. Academy.* B 50, 242-244.

Madhurima Dikshit and Dikshit, M. 1992. Effect of temperature on the ascorbic acid contents of sprouting *Dolichus biflorus*. *Indian J. Nutr. and Dietetics* 29: 254-260.

Murthy, N.R.S., Srinivasa, and Murthy, N.R. 1980. Rapid isolation of high molecular weight plant DNA. *Nucleic Acid Research*. 8: 4321-4325.

Mahadevappa, V.G. and P.L. Raina 1978. Nature of some Indian Legume. lipids. *J. Agril. and Food Chem.* 26: 1241-1243.

Mukherjee, T. (1984). *Indian Drugs*. 21, 224.

Mandal, S., Sardana, S. Negi, K.S., Vasta, P., Phogat, B.S. and Singh, M. 2004. Quality traits of arid legumes with special reference to wild horsegram (*Macrotyloma sar garlwalensis*) IC 212722. *National symposium on Arid legumes for sustainable agriculture and trade,* CAZRI, Jodhpur, Nov. 5-7, 2004 pp. 46.

Omokanye, A.T. 1995. Effect of seed coat colour on seedling and other plant characteristics in horsegram. *Seed Res.* 23: 22-24.

Peshin A. and Singhla, S.K. 1994. Anticalcifying properties of *Dolichos biflorus* (horsegram) seeds. *Indian J. Experimental Biology*. 32 (12): 889-891.

Pore, M.S. 1979b. Mineral composition and oxalic acid in horsegram. *Indian J. Agric. Sci.* 49: 712-714.

Ray, P.K. 1969. Toxic factors in raw horsegram. *Journal of Food Sci. and Techno.* 6: 207-208.

Ray, P.K. 1970. Nutritive value of horsegram III. Determination of biological value, digestibility and net protein utilization. *The Indian J. of Nutr. and Dietetics* 7: 71-73.

Satpute, M.R., D.M. Gupta and S.S. Chavan 1996. Nutritive value of horsegram. Mutants. *J. of Maharashtra Agril. Univ*. 21: 210-212.

Sudha, N., J.M. Begam, K.P. Shambulingappa and C.K. Babu 1995. Nutrients and some anti-nutrients in horsegram. *Indian J. Expt. Biol.* 26: 618-620.

Satpure, M.R., Gupta, D.N. and Chavan, A.S. 1995. Mineral composition of horsegram mutants. *J. Maharastra Agric. Univ.,* 20 (3): 455-456.

Singla, S.K. and Kumar, K. 1985. Inhibitors of calcification from *Dolichus biflorus. Proceed Third Nat. Conf. of Uralithiasis, Soc. of India,* Udaipur, 27-28, February, 51.

Solbrig, H. 1983. Anatomical studies on *Dolichos biflorus L.* with special reference to the ontogeny of calcium oxalate crystals in the seeds. *Dissertation Fachbereich Biologie Universitat*, Hamburg, p. 119.

Thakur, 1979. Scientific Crop Production Vol. 1. *Cereals and pulses* Metropolitan Book Co. Pvt. Ltd., New Delhi-110 002.

7

Horsegram in Uttaranchal – Status, Constraints and Prospects

S.K. Shukla, V. Mahajan and H. S. Gupta

STATE GEOGRAPHY

Uttaranchal (UA), the 27th state of India carved out from erstwhile Uttar Pradesh, is the 10th Himalayan state of the country. It extends from 28° 43' N to 31° 27' N latitude and 77°34' E to 81°02' E longitude. Uttaranchal is surrounded by Nepal in the East, China in the North, Himachal Pradesh in the west and U.P. in the South. The hilly region of Uttaranchal can be classified into four subzones, namely, the low-hills (600-1000 m), mid-hills (1000-1500 m), high hills (1500-2500 m), and very high hills (2500-8000 m). In the low hills, the climate is sub-tropical to warm temperate; in mid-hills, wet and temperate; in high hills cool and temperate; and in very high hills, there is permanent snow cover in area at 5000 m and above. The annual rainfall ranges from 1000 to 2000 mm and temperature varies from less than zero in higher hills to 35^{0}C in plains.

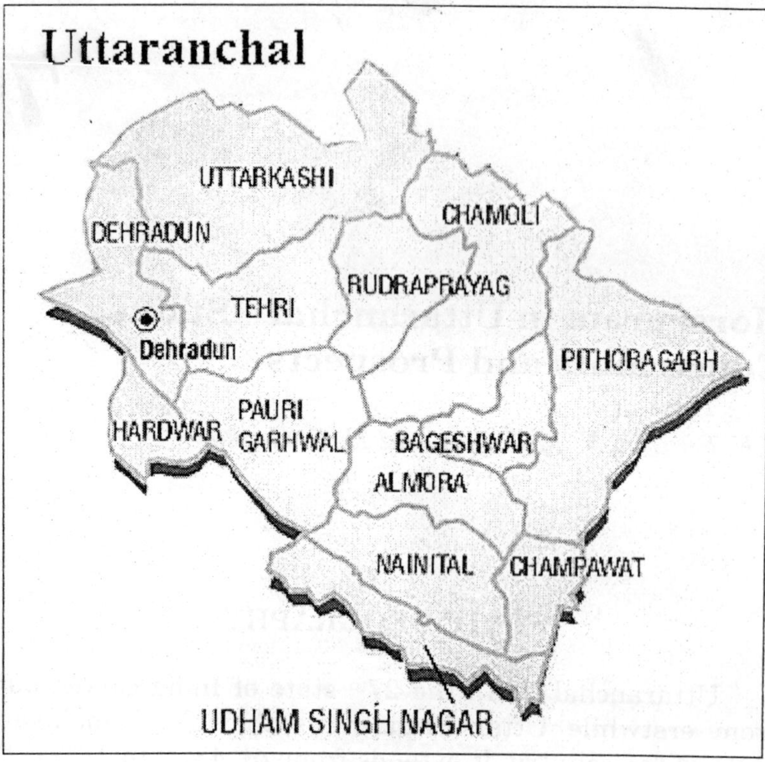

Map of Uttaranchal

Total geographic area of Uttaranchal is 53.48 lakh ha and only 7.93 lakh ha (3.39 and 4.45 lakh ha in plains and hills, respectively) is suitable for cultivation. In plains, 2.97 lakh ha (88.0%) is irrigated, however, only 0.50 lakh ha (11.0%) is irrigated in hills. Hilly areas of Uttaranchal can be divided into two distinct geographical regions viz., Kumaon and Garhwal regions. Kumaon region comprised of six districts viz., Nainital, Udham Singh Nagar, Almora, Bageshwar, Pithoragarh, Champawat, whereas Garhwal regions has seven districts viz., Dehradun, Paudi, Tehri, Chamoli, Rudraprayag, Uttarkashi and Hardwar (Fig. 1). Out of these 13 districts, Udham Singh Nagar and Hardwar fall in plains and remaining 11 are characterized as hilly districts. In Uttaranchal, soil and irrigation facilities considerably vary from place to place. The two districts in plains have good natural resources to grow irrigated rice and wheat while in 11 hilly districts, 89 per cent area is rainfed and only 11 per cent area is irrigated and mostly located in valleys. Land use pattern in Uttaranchal is given in Table 1 & Fig. 1.

Table 1. Land use pattern in Uttaranchal

S. N.	Particular	Area (000' ha)
1	Forest cover	3466
2	Culturable waste	323
3	Fallow	083
4	Barren and unculturable lands	295
5	Land put to un-agricultural uses	167
6	permanent pasture and other grazing lands	223
7	Misc. tree crops and groves	216
8	Net sown area	793

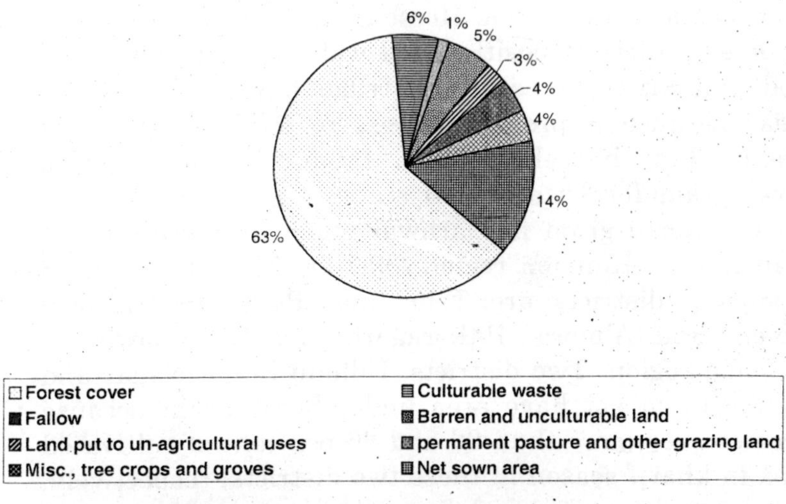

Fig. 1. Land use pattern in Uttaranchal

HORSE GRAM

Horse gram (*Macrotyloma uniflorum* [Lam.] Verde.) is a branched, trailing or sub-erect annual pulse crop. It is known by different names in different languages, viz., Kulaththa (Sanskrit), Kulthi, Gahat (Hindi), Kurtikalai (Bengali), Kulthi (Marathi), Kalathi, Kulit (Gujrati), Muthira (Malyalam), Ulavalu (Telgu), Lollu (Tamil) and Hurali (Kannada). Hymowitz (1990) classified legumes under three categories, viz., primary, secondary and tertiary legumes. Horsegram comes under tertiary legumes. Tertiary grain legumes are those legume species rarely grown outside their native habitat. They are not true domesticated plants but rather cared for or protected plants whose seeds are harvested by the indigenous populations primarily during time of environmental stress. The U.S. National Academy of Sciences has identified this legume as a potential food source for the future (NAS, 1978).

AREA, PRODUCTION AND PRODUCTIVITY

Horse gram is one of the most important kharif pulse crops of the Uttaranchal. Horse gram as traditional crop occupies a distinct position and well defined role in the food and nutritional security of tribal, backward and hilly areas because it provides access to sufficient food for livelihood at household level. Production statistics of horse gram for Uttaranchal state is given in Table 2. Area of horse gram in Garhwal region is more (60.3%) than that in Kumaon region (39.7%). Major horse gram producing districts are Tehri and Paudi in Garhwal region, and Almora, Pithoragarh and Champawat in Kumaon region. Two districts, Udham Singh Nagar and Haridwar do not have area under horse gram because rice occupies more than 96 and 86 per cent of cultivated area in kharif season in these two districts, respectively. Uttaranchal is a pulse deficit state as it produces only

0.30 lakh tonnes of pulses which is only 20 per cent of its total requirement of 1.54 lakh tonnes (Anonymous, 2004) (Table 3).

Table 2. Area, production and productivity of horse gram in Uttaranchal

Period	Area (ha)	Production (metric tons)	Productivity (kg/ha)
2002-03	11,334	6801	600
2003-04	10,807	7953	732
2004-05*	11,470	9070	791

Source: Kharif Abhiyan, 2004, * Targeted

Table 3. Pulse area, production, requirement, productivity and sufficiency/ deficiency in UA, NW Hills and India (2004)

	Area ('000 ha)	Produ-ction ('000 t)	Requir-ement ('000 t)	Suffi-ciency/ deficiency ('000 t)	Average produc-tivity (kg/ha)
Uttaranchal	46.0	31.0	154.7	-123.7	674
NW Hills	128.5	81.8	449.4	-367.6	635
India	20049.6	11143.1	18,304	-7160.9	556

PRODUCTION PARAMETERS

Horse gram is grown as main kharif pulse crop in hills of Uttaranchal. Other pulses grown during kharif season are urd, bhat (black soybean) and rajmash. Generally, horse gram is grown as mixed crop with finger millet (*Eleusine coracana* (L.) Gaertn.). Farmers prefer mix crop of horse gram with finger millet to avoid rotting of stem, leaf and pods in rainy season due to its indeterminate nature on soil surface. Finger millet gives support to horse gram plant. As finger millet matures earlier, farmers harvest the panicles of finger millet and leave its stem for support to horse gram crop.

Being a leguminous crop, it also fixes the atmospheric nitrogen in the soil. Besides, its dense foliage prevent the soil erosion at slope of the hill terraces during heavy rains. Moreover, it withstands drought and survives even in very poor soils. Low income group uses it as main source of protein. It also possesses medicinal property and is very useful in curing the problem of kidney stone, which is common in hills due to high mineral content in drinking water. The popular recipes of horse gram in Uttaranchal are dubka, chhillad and parothe. Besides, dry straw of horse gram is very nutritive feed for milch animals. These attributes make it one of the most potent dual purpose pulse crops of Uttaranchal hills.

In hills, horse gram is sown during month of June and harvested in October. Some of the major diseases of horse gram in hills are Anthracnose (*Colletotrichum lindemuthianum*), leaf blight (*Rhizoctonia bataticola* or *Macrophomina phaseolina*) and white rot (*Sclerotinia scleroretiorum*). Some of the insects that attack horse gram are sucking bug (*Choliops choprai*), girdle beetle (*Oberia* sp.), blister beetle (*Mylabris palarata*) and hairy caterpillar (*Spilartia oblique*).

VPKAS, ALMORA – A lead centre in horse gram research in NW Hills

After realizing its importance in hills of Uttaranchal, horse gram was included as one of the mandated crops of the Vivekananda Parvatiya Krishi Anusandhan Sansthan (VPKAS) way back in 70's. In beginning, a large number of germplasms were collected from different parts of Uttaranchal hills. Thereafter, they were evaluated for yield and further selected. VL Gahat 1 was one of such selections and released in 1983. VL Gahat 1 has become very popular among hill farmers. It generally matures in about 130-135 days, but the same variety matured in less than 90 to 95 days at Ranchi, Bihar (Manigopa Chakraborty, personnel communication).

Some characteristic features of VL Gahat 1 at Hawalbagh Farm, Almora, UA are as below:

Plant height	55-60 cm
Maturity duration	130-135 days
Seed-coat colour	Bright yellowish brown
Yield potential	1200 to 1500 kg/ha (in Uttaranchal hills)
Reaction to diseases	Resistant to leaf blight

At Present, 141 accessions of horse gram are being maintained at VPKAS, Almora. Some of the promising accessions identified for different traits are given below:

Trait	Promising accession
Pod length > 5.9 cm	IC 255434
Seeds/pod >6.5	IC 316167, IC 281677
Pods/plant 100	IC 393211, IC 355837, IC 281677, IC 317400, IC 279506
100-grain weight > 3.4g	IC 316185, IC 281656

(Source: Anonymous, 2005)

Vigorous hybridization program was initiated in 1994-95 and segregating generations were advanced through pedigree method. This approach resulted into some lines *viz.,* VLG 8, VLG 10, VLG 15, VLG 16 and many others. Among them, VLG 8, VLG 10 and VLG 15 have already been identified for release for Uttaranchal hills. Besides, VLG 8 and VLG 10 are being tested in All India Coordinated Program and found to be promising in Initial Varietal Trial (IVT) during kharif 2004 across 8 locations (Almora, Bawal, Bhilwara, Nadia, Pratapgarh, Ranchi, S.K. Nagar and Solapur) in Northern zone (Fig. 2).

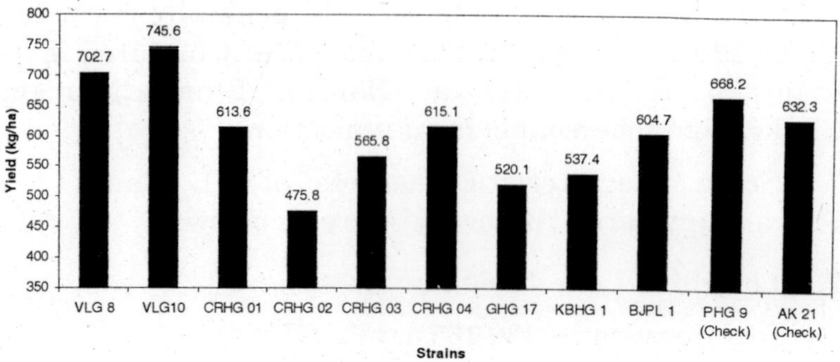

Fig. 2. Mean performance of horse gram strains in IVT during kharif 2004 in Northern Zone (over 8 locations)

VLG 10 showed multiple disease resistant reaction (<10% infection) against root rot, powdery mildew, anthracnose and leaf blight (CAZRI, 2004-05). Besides, VLG 10 also had resistant against red spider mites (Table 4).

Table 4. Reaction of VLG 10 against major diseases and insect-pests (kharif 2004)

	VLG 10	AK21 (Check)	PHG9 (Check)
Disease			
Root rot complex (%)	8.2	6.8	3.4
Powdery mildew (1-5 scale)	1	1	1
Leaf blight (%)	4.7	17.2	6.8
Anthracnose (1-9 scale)	3	7	9
Insect			
Mites/3 leaves	2.30 (4.79*)	4.33 (18.25)	3.23 (4.88)

* values in parentheses are transformed values

Quality attributes

Quality parameters of some of the promising genotypes were assessed along with check VL Gahat 1. Details are given in following table:

Parameter	VLG 1(c)	VLG 8	VLG10	VLG14	VLG15
Protein content (%)	28.9	26.2	25.4	26.7	25.4
Density (g/ml)	1.17	1.10	1.26	0.91	1.26
Volume expansion (%) 24hr	140.0	133.3	160.0	85.7	140.0
Solid expansion(%)	22.32	26.78	19.02	23.3	23.97
Water absorption (%) 24hr	92.3	101.1	110.2	99.0	99.1
Cooking time (min)	35	35	34	25	34
100-grain weight (g)	2.94	3.31	3.15	3.20	3.14

(Source: PAU, Ludhiana)

Under organic conditions, performance of horse gram was also very encouraging. In Uttaranchal, State Varietal Trial (SVT) of horse gram during 2004 was conducted under organic mode only over six locations, viz, Hawalbagh (Almora), Majhera, Sariapani, Thal, Majhkhali and Chinyalisor. Mean grain yield of seven test entries along with check VL Gahat 1 at Hawalbagh centre is given in Fig. 3.

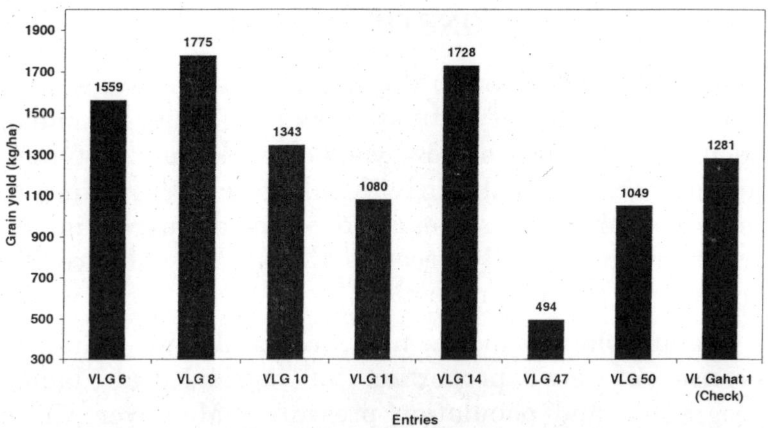

Fig. 3. Mean grain yield(kg/ha) of seven test entries along with check VL Gahat 1 at Hawalbagh (Kharif 2004) under organic conditions

To make this crop more popular and highly remunerative among farmers, 159 frontline demonstrations (FLDs) of VL Gahat 1 were conducted during 2003 by VPKAS, Almora. Average yield of VL Gahat 1 was found to be 532 q/ha as against local varieties (397 q/ha) with a yield superiority of 34 per cent (Gupta *et al.*, 2003)

New initiatives

1. Farmer's Participatory Breeding program has been started in horse gram during kharif 2005. In this program, advance lines of horse gram in station trials are exposed to 15 progressive farmers representing major horse gram growing areas in Uttaranchal. Genotypes are selected by farmers through two approaches, i.e., sequential and comparative. In sequential approach, farmers first see each entry one by one and tell about desirable and undesirable features of each and every genotype. Thereafter, in comparative approach, farmers revisit all the selected genotypes and sort out top three genotypes.
2. Characterization of micro-symbioant in horse gram to assess the nodulation process

CONSTRAINTS

In Uttaranchal hills, the use of local varieties with traditional method of cultivation by the farmers is one of the major reasons for low yield. Development of high yielding varieties with fairly high degree of resistance to biotic and abiotic stresses, and suited to agro-climatic conditions in the hilly areas is the key to enhance the productivity as well production.

Land holdings in the hills are small and scattered because of low proportion of agricultural lands, topography, and population pressures. Moreover, there are large variations in soil and land productivity and irrigation potential. In Uttaranchal, 71 per cent holdings are smaller than 1 ha and 16 per cent holdings are between 1.0-2.0 ha (Table 5). Thus, 87 per cent of the holdings in Uttaranchal are not economically viable, and it is challenging to apply any improved technology to generate sizeable income from such small parcels of land.

Table 5. Operational holdings in Uttaranchal

S. No.	Holding type	Unit	Area (ha)
1	Marginal (<1.0 ha)	6,64,000	98,000
2	Small (1.0-2.0 ha)	1,53,000	3,49,000
3	Large (>2.0 ha)	1,10,000	4,12,000

Besides, there are a number of other constraints to the productivity of pulses in general and horse gram, in a particular, in Uttaranchal hills:

1. Unavailability of high yielding, short duration varieties with resistance to biotic and abiotic stresses is one of the most potent bottlenecks in increasing the production of horsegram in hilly areas of Uttaranchal.

2. Horse gram is generally grown on marginal lands. horse gram is very sensitive to poor water regime, micro-nutritional regime and poor soil physical conditions which affect germination, plant growth and seed yield.

3. Broadcast method of sowing leads to uneven plant stand and creates problems for adopting essential agronomic practices such as weeding and inter-culturing. Lack of efficient *rhizobium* cultures is an area of concern that affects growth and yield.

4. Lack of seed availability of improved varieties of horse gram is a big constraint for increasing its production. The seed replacement rate is very low; it is about 2.0 per cent against a target of 10 per

cent. The farmers are thus compelled to grow land races, which are poor yielders.

5. A committed transfer of technology programme is essential for achieving the goals in pulse production. The extension, training and human resource development components have to be re-energized for achieving self-sufficiency in horse gram.
6. Post harvest handling, processing and non-availability of proper storage facilities with farmers and the vulnerability of pulses, particularly to storage pests cause tremendous losses. Non-processed pulses can not be stored for longer duration. Small scale processing units at the village level will meet the local requirements effectively.
7. Prioritization of market need-based cultivation of selected pulses are either lacking or insufficient. Assured procurement and adequate price for all the pulses will go a long way in promoting pulses farming.
8. Land use is generally not in line with land capability. Additions to cropped area have come from marginal, sloping lands with poor soil where productivity gains are most difficult to achieve.
9. Non-adoption of soil and water conservation measures, sound agronomic and land husbanding practices, and improved cropping pattern.
10. Higher cost of production.
11. Migration of labours to cities has in many cases left old people and women to practice farming.

THRUST AREAS

1. Most of the breeding programme for improved yields in pulses including this crop rely on

empirical selection for superior seed yield and quality across a wide range of target environments. While direct selection for seed yield can be effective (White *et al.*, 1994), the approach is difficult and costly, and gains from the selection are often low in stress prone environments. Inheritance or repeatability of seed yield is very low, which is indicative that the observed variation that is attributable to genetic effects (G) is relatively small in compar-ison to variations observed due to environmental effects (E) and that just as different grain legume species respond differentially to different environment, so different genotypes respond differentially depending on environment (GxE) (Williams, 1992; Blum, 1988). In Uttaranchal, agro-climatic variat-ions are quite significant due to altitudinal difference and sun shine hours. In such areas the genotype x environment interactions are very important determinants of crop yields. In view of above considerations, multilocation evaluation of horse gram genotypes over years under both favourable and stress prone environments is needed to identify genotypes suitable for different target environments and to get maximum gain from selection for grain yield and other yield contributing traits.

2. Availability of large-scale plant biodiversity in Uttaranchal is a great wealth, much of which has yet to be evaluated and exploited. Inclusion of promising accessions in breeding program and incorporation of useful agronomic traits like resistance to biotic and abiotic stresses from promising accessions into new varieties of horse gram will be very effective.

3. Farmers generally cultivate land races which are suited to local conditions. The spread and adoption

of improved varieties of pulses in these areas is very meager because of the unavailability of sufficient quantity of quality seeds. A vigorous seed production programme coupled with efficient transfer of technology for fast dissemination of new and improved varieties is very necessary.

4. There is ample scope for increasing the area under pulses in the hills. This can be achieved by suitable crop intensification, intercropping and modifying the cropping system. The horizontal increase in area is possible through diversion of area from cereals to pulses. Intercropping of pulses with cereals is a better alternative for bringing more area under pulses without much sacrificing the area under cereals. At this stage the role of crop management becomes important to fulfill the requirements of the paradigm: genotype x environment x management. Productivity improvement with combination of high yielding varieties and suitable management practices is essential.

5. Horse gram, as a cheap source of protein, has very good prospects in hills of Uttaranchal. Pursuance of intensive breeding program for the development of high yielding varieties with good cooking quality is area of concern. Medium-tall, semi-erect, early duration, higher poding with bold seed size and cold tolerance are some important traits for development of high yielding horse gram varieties. Proper evaluation, characterization and utilization of vast genetic variability for horsegram in NW Himalayan region will be very helpful in genetic improvement of this crop.

References

Anonymous, 2005. Final Report, Jai Vigyan – National Science and Technology Mission on *Conservation of Agro-Biodiversity*, VPKAS, Almora.

Anonymous. 2004. *Fertilizer Statistics 2003-2004.* The Fertilization Association of India, New Delhi.

Blum, A. 1998. *Plant Breeding for Stress Environments.* CRC Press, Boca Raton. FL.

CAZRI (Central Arid Zone Research Institute). 2004. *Annual Progress Report 2004-05 of National Network Research Project on Arid Legumes.* Jodhpur, Rajasthan.

Gupta A., Gupta H.S., Singh B.D. and Chaudhary B. 2003. Frontline demonstration for household food and nutritional security for tribal, backward and hill areas through traditional crop species in Uttaranchal. *VPKAS Newsl.*, 7(2): 1-2.

Hymowitz, T. 1990. Grain legumes. P 54-57. In: J. Janick and J.E. Simon (Eds.), *Adv. in New Crops.* Timber Press, Portland, OR.

Kharif Abhiyan. 2004. *Agriculture Department,* Govt. of Uttaranchal (Hindi). pp 41.

NAS (National Academy of Science). 1978. Tropical Legumes: Resources for the future, National Academy of Sciences, Washington, D.C., 75 pp.

White, J.W., Ochoam, R., Ibarrap, F. and Singh, S.P. 1994. Inheritance of seed yield, maturity and seed weight of common bean (*Phaseolus vulgaris*) under semi-arid rainfed conditions. *J. Agric. Sci.,* 122: 265-273.

Williams, J.H. 1992. Concepts for the application of crop physiological models to crop breeding. In: Nigam, S.N. (Ed.) Groundnut - A Global Perspective. ICRISAT, Patencheru, India.

8

Prospects of Horse gram (*Macrotyloma uniflorum* Lam. Verdc.) Cultivation in Gujarat State

S.B.S. Tikka, R.M. Chauhan, S.D. Solanki and N.B. Patel

Horsegram [*Macrotyloma uniflorum* Lam. Verde.], locally known as *Kulthi*, *Kalath* in Gujarat is a minor pulse grown in small pockets by the tribals. The total area under its cultivation in Gujarat is around 2160 ha, with average productivity of 526 kg/ha. The major area is confined to Dangs, Surendranagar, Dahod, Panchmahal and Rajkot district (Anonymous, 2000). It is a hardy grain legume with ability to withstand prolonged droughts. The incidence of insects, pests and diseases is almost negligible except red mites which sometimes play havoc and completely destroy the crop. It shows good adaptability to wide range of soil and climatic conditions. It thrives well on well drained soils, particularly hilly slopes. The crop does not require much expensive care and management for its cultivation. People use whole boiled seeds as *dal* which is much preferred by tribals. It is a cheapest source of dietary protein, vitamins, calcium and iron. The crop can be stored for a long time as the seeds are not much damaged by store grain pests. Horse gram has many favourable features that make it apt crop for cultivation in summer such as short duration, ability to withstand prolonged drought, wide adaptability, low incidence of pests and diseases etc.

39 genotypes of horsegram were evaluated at Sardarkrushinagar for protein content which ranged from 17.63% to 23.64%. Highest protein content was recorded in AK-21 (23.64%) followed by GHG-40 (23.61%) and GHG-38 (23.40%) (Anonymous, 2004). Mandal *et al.* (2004) reported quality traits of one wild horsegram accession (IC212722) collected from remote village Sar of Garhwal Himalaya in Uttaranchal. It has registered very high seed protein value of 38.35%.

Varietal Evaluation

Under AICRP programme on Arid Legumes, the varieties were evaluated in Coordinated trials at two location viz., Sardarkrushinagar and Dahod. A SDAU-GVT collaborative programme is in operation in participatory varietal evaluation in horsegram wherein varieties were evaluated in mother trials and baby trials.

Table 1. Evaluation of promising horse gram strains in co-ordinated trial in Gujarat (grain yield kg ha^{-1})

Strains	*Kharif* 2000		*Kharif* 2001		*Kharif* 2002		Mean yield (kg/ha)
	S.K.Nagar	Dahod	S.K.Nagar	Dahod	S.K.Nagar	Dahod	
AK-21(C)	394	692	255	1222	486	551	600
AK-1	396	731	357	1157	483	702	638
AK-42	350	903	504	1333	474	827	732
AK-22	-	-	376	1119	482	679	662
AK-44	-	-	231	931	367	756	571

Coordinated trials : During *kharif* 2000 - 2002, advanced varietal trials were conducted. The results are provided in Table 1. Amongst five varieties evaluated in six trials during three years variety AK-42 registered the highest mean yield of 732 kg/ha which was 22 per cent higher over the released variety AK-21.

Participatory Varietal Selection (PVS) in horsegram

Generally local land races are under cultivation by the tribals which are low yielding, long duration with poor plant type. One way of increasing the spread and adaptation of new varieties for farmers is to give a wide range of novel cultivars to test for themselves in their own fields. The method is referred as participatory varietal selection (PVS).

A successful participatory varietal selection programme has four phases :

1. Participatory evaluation to identify farmers' need in a cultivar.
2. A search for suitable material to test with farmers.
3. Experimentation on its acceptability in farmers fields.
4. Wider dissemination of farmers preferred cultivars.

The cultivars be selected carefully to save time and ensure availability of seed. We have used advanced breeding lines and released varieties for evaluation by the farmers in mother and baby trials. The mother trials were also simultaneously conducted on research farms.

Mother trials : At farmers fields, single replicate designs (mother trial) are used to assess the relative performance of varieties. The trials are researchers design but farmers managed, and they are replicated across village. They not only serve as demonstrational plots or focal points for discussion but are also specifically designed to provide quantitative, analyzable data on yield and yield attributing characters. Six mother trials each during *kharif* 2003 and *kharif* 2004 were conducted in Dahod. The results of these trials along with mother trials on research station are given in Table 2. All the varieties were superior over local check. AK-42 registered 22 per cent higher yield over the released variety AK-21.

Table 2. Grain yield (kg/ha) of different varieties of horsegram in mother trials in Gujarat state (*Kharif* 2003 and *Kharif* 2004).

Varieties	*Kharif* 2003				*Kharif* 2004		
	Grain yield (kg/ha)						
	S K Nagar (NF)	Dahod (NF)	S K Nagar (HF)	Dahod (Farmer's field)	Dahod (NF)	Dahod (Farmer's field)	Mean yield
AK-1	721	704	787	363	-	-	644
AK-21	752	815	714	499	432	391	601
AK-22	779	1100	814	431	-	-	781
AK-42	895	1153	671	563	540	563	731
AK-44	752	882	833	442	509	423	640
Maru K – 1	656	782	683	459	432	471	581
Local	490	500	413	316	-	-	430

NF = Normal fertility HF = High fertility

Baby Trials : Baby trials are "Farmer Managed Participatory Research Varietal Trials". In Baby trials only farmers' perception on yield and other characters is collected. The farmers grow one or two new varieties along side the local variety and farmers managed practices are given. Replication is across villages. The trials are evaluated by participatory methods like farm walks, focus group discussion and house hold level questionnaires. Twenty eight Baby trials during *kharif*

2003 and 27 trials during *kharif* 2004 were conducted in four different villages of Dahod district. For monitoring of trials frequent visits were conducted at different stages of the crop growth and perception of the farmers was taken. Results are given in Table 3 and Table 4.

Table 3. Perception of Baby trials (*Kharif* 2003)

Sr. No.	Perception	Number of farmers			
		V_1 (AK-42)	V_2 (AK-21)	$V_1 = V_2$	Local
1.	Yield	19	8	1	0
2.	Earliness	23	0	5	0
3.	Quality	12	0	16	0
4.	Overall preference	18	3	7	0
5.	Better than local	28	28	-	-

Table 4. Perception of Baby trials (*Kharif* 2004)

Sr. No.	Perception	Number of farmers			
		V_1 (AK-42)	V_2 (Maru K -1)	$V_1 = V_2$	Local
1.	Yield	12	7	5	3
2.	Earliness	9	7	7	4
3.	Quality	7	5	10	5
4.	Overall preference	6	8	10	3
5.	Better than local	24	24	-	3

During *kharif* 2003, AK-42 excelled in all perceptions i.e. out of 28 baby trials farmers perception in respect of better than local (28), earliness (23), yield (19), overall preference (18) and quality (12) as compared to AK-21 and local. During *kharif* 2004 also AK-42 excelled in all perceptions i.e. better than local (24), yield (12), earliness (9) and quality (7). In respect of overall preference AK-42 (6) and Maru K -1 (8) were at par.

Thus based on the results of varietal evaluation in coordinated trials, mother trials and baby trials, the variety AK-42 with 22 per cent superiority in grain yield over the released variety AK-21, as well as better consumer preference as reflected by farmers perception was identified for recommending in the horsegram growing pockets of Gujarat state. This variety also had appreciably higher protein content (21.02%).

Focus Group Discussions (FGDs)

Before and after harvest of the trials, FGDs were carried out and perception of the farmers was recorded on various aspects like taste, market price and storability. A post harvest FGDs was carried out in village Jada (District : Dahod). From FGDs following feed backs were obtained :

- Light seed coat colour varieties were preferred by the farmers due to their easy cooking.
- Horsegram was used to check soil erosion on hill slopes.
- Horsegram fodder and grain were used by the farmer for feeding their animals.

Production Constraints

The low productivity of horsegram in the state is attributed to a number of production constraints as below:

1. The crop is most cultivated on marginal lands, shallow soils, hill slopes as rainfed crop.
2. The crop is generally sown as mixed, inter and border with maize or pearl millet during *kharif* season.
3. Lack of high yielding varieties.

4. Lack of fertilizer responsive varieties and genetic variability.
5. Poor adaptation of available technical knowledge.

Future Research Needs

1. Evolving early maturing (80-90 days), photo and thermo- insensitive varieties with high grain yield potential (8-9 q/ha).
2. Evolving suitable high yielding varieties with determinate growth habit for intercropping, particularly, with maize.
3. Introgression of high protein to existing land races from high protein profile.
4. Identification of varieties having high fatty acid particularly linoleic and lenolenic acid which are responsible for lowering cholesterol and solving urinary and kidney problems.

References

Anonymous, 2000. Joint director of Agricultural statistics and economic cell. Government of Gujarat, Krushibhawan, Sector 10-B, Gandhinagar.

Anonymous, 2004. A SDAU-GVT collaborative research project on horsegram and blackgram APR.

Mandal, S., Sardana, S., Negi, K.S., Vasta, P., Phogat, B.S. and Singh, M. 2004. Quality traits of arid legumes with special references to wild horsegram (IC 212722). *Nat. Symp. on Arid Legumes for Sustainable Agric. and Trade.* CAZRI, Jodhpur, Nov., 5-7 2004, pp 46.

9

Present Status and Potential of Kulthi Cultivation with Reference to Himachal Pradesh

H.L. Thakur, Daisy Basandrai, Ashwani K. Basandrai and S.K. Sharma

The present per capita available 1500 m^3 water is shrinking to touch the scarcity line of 1000 m^3 mark sooner or later. It is expected to decline further by 40-60% in the next three decades. The accelerated pace of industrialization and urbanization will further constrain the water resources available to agriculture. The pulses have better water use efficiency on harvested yield (0.2 to 0.6 kg/m^3), so these can be grown under limited water availability and would be preferred over other crops. Approximately 90% of the pulse production comes from rainfed ecosystems. Pulses are the most suited crop for the Himachal Pradesh conditions where, 80% area is rainfed. Among all kharif pulses, urd bean ranks first in Himachal Pradesh followed by horse gram (*Dolichos biflorus* (*Macrotyloma uniflorum* (Lam.) Verde).

Horse gram is grown in Eastern Africa, Southern Africa, Malaya, Mauritius and India. In India it is grown in an area of 19.78 lakh ha with annual production of 6.66 lakh tones. Nearly 90% of the area is distributed in the southern states viz. Andhra Pradesh, Karnataka and

Tamil Naidu and rest is in Maharashtra, West Bengal, Orissa, foot hills of Uttranchal, Bihar and Himachal Pradesh. It can be grown over a wide range of soils except on alkaline soils and can also be grown successfully as a preparatory crop on reclaimed lands. In Himachal Pradesh, it is grown in an area of 3348 ha with production of 1360 tonnes and productivity of 405 kg/ha. It is grown as a monocrop and as an inter crop with cereals, groundnut, sesame etc. The seeds are rich in protein (22%), carbohydrates (57%) and minerals. It is a good source of lysine but deficient in sulphur contai-ning amino acids. The kulthi seeds are not very palatable and tend to have an earthy aroma but its daal is good for patients suffering from kidney stone problem (Duke, 1981). The seeds are used in preparation of delicious dish 'Kuthicut'. It can also be used as forage and green manure. It furnishes concentrated feed for cattle, especially for horses.

CROP MANAGEMENT

(a) Soil

It is generally grown on marginal lands. Well drained sandy/ clay soils are suitable for its cultivation. There is no problem of kulthi cultivation in Himachal Pradesh as the land in hills is generally sloppy and there is no problem of water stagnation. Land is prepared by giving two to three ploughings followed by planking.

(b) Sowing time and seed rate

This crop is sown with the onset of monsoon by the last week of June or beginning of July. The sowing is done in lines 30 cm apart by *kera* method. However, farmers generally take the crop by broadcasting method Recommended seed rate is 18-20 Kg/ha. The seed rate is reduced to 1/3rd, when it is intercropped with maize. Practically, farmers use seed rate according to availability of their own seed from the previous crop.

(c) Nutritional requirements

The crop is grown without application of any manure or fertilizer. The recommended nitrogen and phosphatic fertilizer application is 12-15 and 45 kg /ha at the time of sowing respectively. There is no need of nitrogen application when it is grown mixed with maize. Under rainfed conditions, nutrient dose is reduced to half.

(d) Interculture Operations

Two weedings and hoeings are recommended to control the weeds. Pre emergence spray of Lasso 50 WP @ 1.5kg/ha/ stomp @ 3L/ha are also effective in controlling the weeds. If need be, one hand weeding can be done, but farmers do not pay any attention to the interculture operations and it leads to further reduction in yield.

(e) Harvesting

The crop is harvested when the plants turn half yellow and pods start drying. Harvesting is done in the month of November. Due to the lack of interculture operations, weeds grow profusely before harvesting in poorly prepared fields. Many of the weeds mature by this time and their seeds are shed in the field or are mixed with *kulthi* during harvesting. The plants are harvested by pulling out whole plant. They are taken to the threshing floor, stacked for a week for drying and threshed by trampling under the feet of oxen or with wooden sticks.

Horsegram improvement programme in Himachal Pradesh

In Himachal Pradesh, *kulthi* is grown in an area of 3348 ha with production of 1360 tonnes and productivity of 405 kg/ha. In Himachal Pradesh, local varieties or landraces are grown since ages. The information on the area, production and productivity of horse gram growing districts is given in Table 1.

Table 1. District wise area and production of horsegram in Himachal Pradesh during 1997-98.

District	Area(ha)	Production(Tonnes)	Productivity(kg/ha)
Bilaspur	54	30	563
Chamba	334	113	338
Hamirpur	131	63	479
Kangra	390	163	417
Kullu	71	11	148
Mandi	497	491	988
Shimla	773	213	276
Sirmour	439	125	285
Solan	459	117	209
Una	100	34	344
Total	3348	1360	405

Source: H.P. Statistical outline.

Horsegram is generally grown on marginal lands hence, it never attracted the attention of farmers for improved varieties. Two improved varieties HPK-2 and HPK-4 (Baiju) were developed and released for cultivation in Himachal Pradesh. Both the varieties are a pure-line selections from the elite germplasm collection. These varieties are semi-spreading with climbing growth habit and early (maturing in 100-125 days). The pod length in variety HPK-2 is about 5 cm with 5-6 seeds/ pod and pods have shattering tendency. Variety HPK-4 (Baiju) is resistant to shattering and its pod length is 4-5 cm with 4-5 seeds/ pod. Both the varieties show resistance to leaf spot diseases. Variety HPK-2 can withstand drought conditions. The average yield of both the varieties is about 10 q /ha.

Bottlenecks in the production of Horsegram

A number of factors are responsible for the limited spread and production of horsegram in Himachal Pradesh:

(a) Poor yield potential of the varieties/landrace

The low yield potential of the existing local varieties/ landraces under cultivation is the major bottleneck in increasing horse gram production in the State. Most of the land races are of very long duration and there is an urgent need to breed or introduce early maturing varieties and to popularize the recommended varieties.

(b) Poor agronomic practices

The farmers usually grow this crop under poor and marginal field conditions. No manures or fertilizers are applied and no cultural operations are conducted.

(c) Biotic stresses

Among biotic stresses viz., leaf spot diseases caused by *Collectotricum lindemuthianum*,, Cercospora leaf spot caused by *C. dolichi*, aerial leaf spot caused by *Rhizoctonia. solani* and viral diseases are the major maladies causing reduction in the production of horsegram.

Future scope and challenges ahead for Horsegram production

Horse gram is still a preferred crop for its nutritional and medicinal qualities. It requires planned research programme in association with AICRP on Arid Legumes under the aegies of ICAR.

Himachal Pradesh has immense potential in the availability of diverse germplasm comprising of landraces and local varieties adapted to varied agro-climatic conditions. Horse gram like other pulses had been grown by the resource poor farmers, under poor and degraded soils without proper agronomic input. It resulted into selection of varieties/landraces adapted to poor soil and agronomic conditions. Consequently, it resulted in the erosion of genes responsible for high yield and productivity. Nevertheless, the genes for high productivity may be scattered in the landraces. Therefore, collection of horsegram varieties/landraces and identification of genes for high yield is the prerequisite for initiating the breeding work. Therefore, genes for higher productivity can be combined through the use of intensive hybridization programme followed by selection. Besides, useful variability can also be generated through mutagenesis.

Proper crop Management will prove a boon for the production of *kulthi*. Training and orientation camps should be organized at the district and block levels to impart sufficient training to acquaint the farmers with the latest information and techniques of crop management.

Horse gram can play an important role to combat the increasing demand of pulses for ever increasing population through intensive and coordinated research and field application oriented programme at the state and National level.

Future strategies

To help horse gram, realise its full genetic potential as a valuable crop, concerted and dedicated research efforts are required in the following direction:

1. To collect, evaluate, characterize and maintain the germplasm resources.
2. To compare the ecological limit and optimal management factors for crop, both for the monoculture and mixed cropping.

3. To breed genotypes having high harvest index, early maturity, compact type, high yield and showing tolerance/resistance to pests and diseases.
4. To breed types with superior nutritive value richer in proteins and limiting aminoacids.
5. To investigate the nutritional aspect of seeds, biochemical analysis, value addition
6. High diversity build up in horse gram in relation to differences in soil, cropping pattern and climatic conditions indicated the need for conservation of diversity at farmers fields for maintaining and perpetuation in the process of continuing adaptation and evolution. On farm conservation activities have already been initiated for horse gram in tribal areas of eastern and peninsular India under the aegis of ICAR. This effort need to undertaken for the diversity in hilly regions of Himachal Pradesh.
7. Construction of detailed genome map for using molecular markers to locate genes of interest and their precise transfer through marker assisted selection. Need to identify more markers.

REFERENCES

Duke, J.A. 1981. *Handbook of Legumes of world economic importance.* New York, Plenum.

10

Horsegram in Andhra Pradesh

P. Raghu Ram Reddy

INTRODUCTION

The grain and culmns of horsegram are very much valued as cattle feed. The crop is drought tolerant and also an excellent contingent crop for late kharif. The crop is grown in marginal soils and mostly without fertilizers and plant protection. Research work on this crop is also less and hence, the low yields in AP. It grows as a pure crop and can be raised as an inter and sequence cropping also. The crop can fit as an intercrop not only between other crops but as well in Agri-horti systems in various orchards like mango, guava etc, apart from Agri-Silvi-Pastoral systems

Being legume it can also be takenup as an organic source of fertilizer with less requirement of fertilizers to the next crop in the same piece of land. Horse gram as a contingent crop is grown late in kharif and comesup well even under residual moisture conditions. *Sambar* is a delicious item made from horse gram and is considered very nutritious in Andhra Pradesh.

Peninsular India as a center of diversity

Eight Agro-Ecological regions are recognized in India based on physiography, climate, soil and associated factors wherein the entire native diversity in plant genetic resource is distributed viz.,

(i) Western Himalayas

(ii) Indus plains

(iii) Eastern Himalayas

(iv) Eastern peninsular region

(v) North Eastern Himalayas

(vi) Western Peninsular Region

(vii) Gangetic Plains

(viii) Island Regions

Horse gram is grown with wide diversity in two of these regions viz., Eastern Peninsular Region/Eastern Ghats/Deccan Pleateau and Western Peninsular Region/ Western Ghats/Malabar.

AREA, PRODUCTION AND PRODUCTIVITY

The range of area, production and productivity of this crop in AP from 1993-94 to 2001-02 was 70 to 139 ha (000), 20 to 49 tonnes (000) and 257 to 489 kg/ha respectively (Table 1). Interestingly there was a decline in area. The production, was declined less than the area decline, the productivity however, showed no such trend. This might be due to availability of better genotypes coupled with better management over the years although very little work on crop improvement as well as crop management and protection has been done with this crop in the state

The data from 1993-94 to 1998-99 showed that the rabi area is higher in AP than the kharif area in a proportion of 1:4/5 while the rabi production was higher than kharif in a proportion of 1(kharif) : 3 to 6(rabi). The rabi productivity was however similar to that of kharif with the kharif and rabi yield/ha being in the ratio of 1 : 0.7 to 1 : 1.6 with similar trend in most of the years.

In kharif season, the Telangana area was highest ($2/3^{rd}$) followed by Rayalaseema and Coastal; while in rabi it was Coastal ($3/4^{th}$) followed by Telangana and Rayalaseema. The region wise production also showed the same trend as that of the area trends mentioned above. Rayalaseema had smaller areas of 5000 to 8000 ha in kharif and 3000 to 6000 ha in rabi with a production of 2000 to 3000 tonnes in kharif and 1000 to 3000 tonnes in rabi.

As per the land utilization (1990-91) of the Directorate of Economics & Statistics, 1990-91, A.P (Table 2.5 & 2.6, Pp 201 of NBSS Pub. 69-Soils of A.P. for Optimising Land Use), out of total land area of 2,74.40.049 ha, the total cropped area was 1,31,92,714 ha.Out of this, horsegram had occupied an area of 1,29,838 ha (1% of the cropped area of the state).The average yield of horsegram (Table 2.7, Pp 21 of NBSS, Pub 69) was 305 kg/ha. However, the suggested land use for minor millets and horse gram along with agro-forestry was 4% of the total cropped area (Fig. 8 - suggested land use - A.P. in NBSS publ. 69).

The land use plan suggested for the state helped in balancing the ecology by growing food, oilseeds, pulses and horticultural crops both as mono crops and in agro-forestry and Agri-horti-silvipastoral systems on marginal and submarginal lands. The area of horse gram as suggested, can be enhanced with better varieties, production and protection technologies and due to its ability to fit in the 4% cropped area, there is a need for the researchers to provide a fillip to this gap now existing which is potential for future.

However, the area from 1,29,838 ha in 1990-91 has declined to 80,339 in 2001-02 but the productivity has increased from 305 to 489 in the corresponding period (Table 1). It can be clearly seen that though area is declining, the productivity gains are evident and therefore, scope exists for horsegram a risk aversion crop to be grown in larger areas of marginal and submarginal lands by choosing latest varieties recommended for the state.

Table 1. Area (000 ha), production (000 tonnes) and productivity (Kg/ha) in Andhra Pradesh State

Year	Total Area				Kharif				Rabi			
	C	R	T	Total	C	R	T	Total	C	R	T	Total
2001-02	53	8	19	80								
2000-01				78								
1999-00				101								
1998-99	71	9	31	111	1	5	13	19	70	4	18	92
1997-98	70	11	29	110	0	6	14	20	70	5	15	90
1996-97	70	9	33	121	1	6	15	22	78	3	18	99
1995-96	83	10	33	126	1	6	14	21	82	4	19	105
1994-95	83	12	39	134	1	6	18	23	82	6	23	111
1993-94	81	14	44	139	1	8	20	29	80	6	24	110
Av				111								
Year	Total Production				Kharif				Rabi			
	C	R	T	Total	C	R	T	Total	C	R	T	Total
2001-02	28	4	7	39								
2000-01				20								
1999-00				36								
1998-99	21	6	9	36	0	3	5	8	21	3	4	28
1997-98	31	4	9	44	0	3	4	7	31	1	5	37
1996-97	32	5	12	49	0	2	6	8	32	3	6	41
1995-96	30	2	12	44	0	2	6	8	30	0	6	36
1994-95	31	4	10	45	0	2	4	6	31	2	6	39
1993-94	23	5	12	40	1	3	6	10	22	2	6	30
Av				39								
Year	Productivity				Kharif				Rabi			
	C	R	T	Total	C	R	T	Total	C	R	T	Total
2001-02	562	460	327	489								
2000-01				257								
1999-00				355								

Year	Total Area				Kharif				Rabi			
	C	R	T	Total	C	R	T	Total	C	R	T	Total
1998-99	379	567	262	321	485	479	377	413	380	655	224	302
1997-98	450	382	281	400	457	459	290	349	450	317	278	411
1996-97	416	600	338	404	474	478	366	398	414	707	310	405
1995-96	391	289	369	351	337	334	389	372	392	245	356	347
1994-95	377	231	272	333	196	196	234	222	379	254	258	349
1993-94	273	368	282	285	399	399	307	334	271	327	261	272
Av				355								

C=Coastal districts; R=Rayalaseema region; T=Telangana region

HORSE GRAM IN INTEGRATED NUTRIENT MANAGEMENT

Long-term field experiments were undertaken on Alfisols of Gungal Research Farm of CRIDA since 1994, with sorghum and sunflower under horse gram incorporation and fallow.

Low organic carbon and poor fertility in soils are generally encountered in rainfed regions. Under erratic and uncertainty of the rainfall, buildup of soil organic carbon (SOC) is low due to poor production of biomass and rapid oxidation from the high temperature and the prevailing moisture. In India the age old practices of using FYM and green manure were in practice for increasing the SOC along with reduction of 'N' requirements. However, the generation of leguminous biomass in a non competitive way in the off season or post rainy season in red soils of semi-arid regions (wherein horse gram is grown effectively as a late kharif crop in residual soil moisture and also utilizes the post-rainy season rainfall accounting to about 20 to 30% of the total rainfall of the year which is 70-100 mm in 7 out of the 10 years) represented by Gunegal Research Farm of CRIDA was found to be a very effective way of soil health improvement. The horse gram was raised in postrainy season and incorporated *in-situ* (Fig. 1) to improve the soil health through enhancement of SOC and reduced the nutrient requerment of the following kharif crops.

Fig. 1: Incorporation of horsegram at pre-flowering stage
Source: CRIDA Annual Report 2003-04. Pp. 19

Horse gram was grown upto flowering in the main plots at site I with sorghum-sunflower rotation and at site II without sunflower-sorghum rotation. The horse gram at flowering stage was incorporated in one plot, while the other was kept as fallow. Various fertilizer treatments were the subplots. The trial was concluded in the year 2004. The results revealed that among soil quality parameters, the organic carbon content was found to be significantly higher under the horse gram incorporation via-a-vis the fallow treatment irrespective of the fertilizer treatments. The Soil organic carbon after 10 years at site I showed 0.28% in fallow for N_0P_0 treatment, while it was 0.37% for horse gram incorporation plot. For $N_{50}P_{30}$ plot, it was 0.46% while in fallow incorporation plot it showed 0.53%. At site II (initial OC was 0.24%), for N_OP_O it was 0.27% while fallow for incorporation plot it was 0.30%. In $N_{50}P_{30}$ plot it was 0.41% for fallow, and 0.52% in the incorporation plot.

The microbial biomass carbon was also 114 mg g^{-1} for N_OP_O in fallow while for incorporation it was 128 mg g^{-1}. For $N_{30}P_{30}$, fallow, it was 217 mg g^{-1} while for incorporation it was 290 mg g^{-1} (Fig. 2). It can be clearly seen that the organic carbon content did not vary significantly. However with horse gram incorporation and N application the available nitrogen status of soils improved up to 21 kg N on an average annually. With horsegram incorporation thr available nutrient status of P, K and S also improved to the extent of 6 kg P, 17 kg K and 7 kg S ha^{-1} annually (added to soil).However continuous application of N and P led to substantial build up of available P, wherein improvement in available N was not significant. Available K increased with incorporation but decreased with the application of $N_{50}P_{30}$.

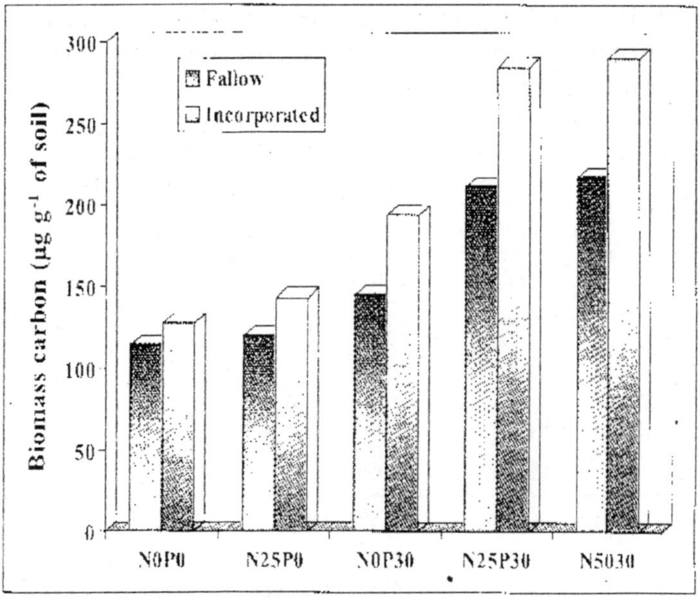

Fig. 2. Effect of long-term horsegram incorporation on soil microbial biomass carbon;
Source: CRIDA Annual Report 2004-05. Pp. 25

The yield data also showed similar trend. Horse gram incorporation and fertilizer treatments significantly influenced the yields of both sorghum as well as sunflower. The treatment $N_{50}P_{30}$ showed highest grain yields in the two crops investigated. Horse gram biomass incorporation resulted in higher yields in these two crops grown in kharif 2004 without any fertilizer treatments for quantifying residual effects of biomass incorporation and nutrient application. Both fertilizer and incorporation treatments enhanced the grain yield significantly (Table 2). The native soil was low in fertility due to deficiencies in available N, S and K as well as low organic carbon content. Horse gram incorporation and fertilizer application both improved the soil quality parameters and nutrient build up in the soil and resulted in higher yields in both crops viz, sorghum and sunflower by 28 and 18%, respectively.

Table 2. Residual effects of horsegram incorporation and nutrient application on grain yield of sorghum and sunflower

Residual fertilizer treatments	Sorghum (grain yield kg ha^{-1})		Sunflower (seed yield kg ha^{-1})	
	Incorporated	Fallow	Incorporated	Fallow
N_0P_0	397	307	343	258
$N_{25}P_0$	758	570	525	436
N_0P_{30}	769	581	536	462
$N_{25}P_{30}$	1040	816	718	637
$N_{50}P_{30}$	1216	990	840	715

Source: CRIDA Annual Report 2004-05 Pp 25

HORSE GRAM AS INTERCROP IN ORCHARDS FOR IMPROVING SOIL FERTILITY AND AS A FODDER CROP

In a study made to investigate the overall sustainability from Horti-Pastoral and Agri-Silvi-Pastoral systems as a need to diversify the existing sorghum production system, it was observed that horse gram could be encou-raged for generation of nutritious fodder for use during the off season (late kharif in AP). The fruit yield of orchards like guava was not negatively influenced by the intercrop upto 8 to 10 years and, hence, can be promoted in orchards this being a leguminous crop also helps in improving the soil fertility. The inclusion of horsegram as an intercrop in between orchards for fodder purpose and inclusion of small ruminants like 'Nellore-Zodpi' (Mahaboobnagar Dist.) and local based (Ranga Reddy Dist) sheep alleviated fodder shortages and added extra income to the fruit based production system.

HORSEGRAM AS A SUITABLE CROP IN FORAGE CROP ROTATION FOR CENTRAL DECCAN REGION

In rainfed and marginal lands of the central deccan region, following crop rotation is recommended by the Regional Station for Forage Production and Demonstration, Mamidipally, Hyderabad (A.P.).

Dina Nath grass + Horsegram-Pillipesara

Following development thrust on the basis of researches and potentialities the following system of forage production have been identified for arid and semi arid regions.

Regions	Accents on
Arid	- Grazing land and silvipasture
	- Hortipasture
Semi-arid	- Integration of agro-silvipasture and dryland agriculture on cultivated lands
	- Forage-cum-coppice farming on the marginal and submarginal lands with intercropping of dryland cereals and legumes

The recommendations revealed that cultivated forage production in the existing area could be intensified with appropriate production technology including improved and recently developed cultivars. An alternative land use system with major thrust on forage resource development should be taken up on watershed and low rainfall/dryland areas. Horsegram can fit in the above recommendations.

CROP IMPROVEMENT

The available local horsegram varieties in the state are low yielding and susceptible to Yellow Mosaic Virus,a major disease in South India including AP.

The varieties released so far for cultivation in the state viz., VZM-1, PDM-1 (developed in the state through the AICPIP in late 60's/early 70's) and Palem-1 are susceptible to Yellow Mosaic Virus. However Palem-2 (1996) developed by Palem Research station of ANGRAU is tolerant to Yellow Mosaic Virus. Details on the same are given below:

PDM-1: This variety was developed by single plant selection from local bulk (a local selection from Bhimilipatnam). It matures in 100-105 days. Plants are spreading type. The flowers are yellow with light purple wash. Seeds are buff coloured. The variety is suitable for kharif sowing and has an yield potential of 1600 kg/ha.

VZM-1: This variety is also developed by single plant selections from local bulks (a local selection from Vizianagaram). It matures in 100-105 days. Plants are tall and semi-erect. Flowers are light yellow with purple wash and seeds are black. It is suitable for kharif planting. Yield potential is 1500 kg/ha.

PALEM-1: This variety was developed at Palem center of Acharya N.G. Ranga Agricultural University. A selection from local germplasm collected from Mahaboobnagar district and recommended through out Andhra Pradesh. It matures in 80-85 days, it is semi-spreading with 35-40 cm tall. The seeds are black in colour. It is suitable for kharif planting. The average yield under normal conditions is 900 kg/ha.

PALEM-2 : Palem center of Acharya N.G. Ranga Agricultural University has developed this variety also.It is recommended for the entire state of AP and was developed by pedigree method of breeding. It is suitable as a rainfed crop in early rabi season. It is semi-spreading with buff coloured and medium size seeds. Matures in 100-105 days. Suitable for both early and normal sowings with seed rate of 20-25 kg/ha. The parentage used were VZM-1 x BGM-1. The average yield under normal conditions is 800-900 kg/ha. This variety is readily accepted by the farmers for its medium bold seed size and buff coloured grains and is gaining popularity in the Southern Telangana Zone of AP.

CRIDA-1-18R : This variety showed an yield advantage of 20.94% over the best check PHG-9. It is also earlier to the check by 9 days. It was superior/on par with regard to quality and disease incidence. Grain yield potential is 842 kg ha^{-1}. The variety is also early maturing in 85 days.

CONSTRAINTS OF GROWING HORSEGRAM IN ANDHRA PRADESH

Horse gram though an important crop of AP, yet occupied 1% of the total cropped area during 1990-91, is declining in area but increasing in productivity. The decline might be due to changing food habits and change in landuse patterns. Also non availability of early and high yielding varieties of horsegram has added to the constraints. Yellow Mosaic Virus is a major problem and horsegram is very much susceptible to this disease due to which yield is affected to a high extent

THE PROSPECTS OF HORSEGRAM IN ANDHRA PRADESH

Horsegram is an excellent organic crop in the real sense. It has high adaptability to come up well under residual soil moisture as well as in erratic and uncertain rainfall prevelant in the arid and semi-arid regions of the state. Its growth with *in-situ* soil and water conservation and other dryland technologies on a watershed basis, is the key of future strategies not only for a higher risk aversion proposition but as the newly developed varieties with inbuilt tolerance to diseases like YMV and powdery mildew etc., are in the making and are recommended for cultivation in the state, the production can definetly be enhanced significantly in a totally organic sense of the term.

In times of drought, the crop has proven a risk aversion crop and has exelled as a contingent crop and shall continue to.

Its role apart from that of a sole crop,as an intercrop and mixed crop in various cropping systems viz., Agro-forestry, Agri-silvi-pastoral etc., is beneficial for soil as well as for animals beings a good source of fodder, especially in arid and semi arid regions of the state.

The description of Agro-climatic zones of AP in the chapter in Integrated Nutrient Management for 2004 revealed horsegram as a major crop along with other crops grown in (II) North Coastal Zone (Srikakulam, Vijayanagaram, Visakhapatnam), (III) Southern zone (Cuddapah, Nellore, Chittoor) and (V) Southern Telangana Zone (Mahaboobngar, Nalgonda, Ranga Reddy, Medak). The chapter on Identification of efficient/ appropriate fodder crops and trees for different AESRs in Andhra Pradesh revealed that horsegram is grown in late kharif for (III) Southern Zone, (IV) Northern Telangana (Adilabad, Karimnagar, Nizamabad, Warangal and parts of Nalgonda, Medak and Khammam) (V) Southern Telangana (Ranga Reddy, Mahaboobnagar, parts of Nalgonda, Warangal and Medak) and (VI), scarce Rainfall Zone (Kurnool, Ananthapur, parts of Prakasam and Cuddapah). It can be seen that out of seven zones, horsegram is mentioned as a major crop/fodder crop in five zones. The other two zones are (I) Krishna Godavari Zone and (VII) High Altitude and Tribal Zone.

However, as per the land use pattern suggested by the National Bureau of Soil Survey and Land Use Planning, Nagpur in cooperation with State Soil Survey, Department of Agriculture, Hyderabad, the minor millets-horsegram based cropping system is a mixture of food crops and agro-forestry. This cropping pattern is recommended for 4.0% of the cultivable land area of AP, with the following primary components.

- Minor millets
- Horsegram
- Custard apple, ber, agro-forestry

The above systems are recommended for very shallow to moderately shallow, mostly red gravelly soils together with rock outcrops, which occur in rolling lands. The districts where such soils and therby the above systems can be successfully practiced are: Ananthapur, Kurnool, Cuddapah, Chittoor, Mahaboobnagar, Nalgonda, Medak, R.R. Dist, Nizamabad, Warangal, Adilabad, Karimnagar and Khammam.

In the above districts with shallow red soils, only short duration kharif crops-horsegram and minor millets can be successfully grown, apart from ber, custard apple and agro-forestry. Horsegram grown with intensive soil and water conservation measures can definetly improve the prospectus of this crop. Four per cent of the total cropped area in the state is recommended for this system and hence, large scope exists for horsegram improvement, production and protection technologies in place.